PREFACE

Since the first report of specific DNA amplification using the polymerase chain reaction (PCR) in 1985, the number of different applications has grown steadily, as have the modifications of the basic method. This increase has been claimed to be exponential - like the PCR itself. In any case, detailed discussion of the full range of biological problems under investigation as well as all of the various experimental approaches utilizing PCR is beyond the scope of this book. Conceived as a snapshot of a rapidly moving field, the book's contributed chapters have been organized into three sections: Basic Methodology, Research Applications, and Medical Applications. I have used the brief introduction to each section to cite some of the more recent literature relevant to each area and to "fill gaps" by directing the reader to topics not covered. Some chapters provide detailed protocols, listing favorite "PCR recipes," while others give an overview of a particular field. My hope is that an interested reader, armed with the knowledge of some of the methodological issues and of some of the applications, could proceed to devise PCR-based approaches relevant to his or her area of investigation.

I am grateful to my current and former colleagues at Cetus for their contributions to this book and, of course, above all, to the technology that occasioned it. To Kary Mullis, who invented the basic PCR method, and to my colleagues in the Human Genetics Department (Randall Saiki, Stephen Scharf, Glenn Horn, Fred Faloona, Kary Mullis, and Norm Arnheim) who first developed it for the analysis of human genetic variation. To David Gelfand and Susanne Stoffel who isolated and characterized the thermostable DNA polymerase (*Taq* polymerase) whose use in the PCR led to the current rapid and automated procedure and transformed the reaction from a method of last resort to one of first choice. To Corey Levenson, Dragan Spasic, and Lauri Goda who responded gracefully to the increased demand for synthetic oligonucleotides generated by the PCR. Also to Tom White who, as Director of Research, saw PCR's potential and supported its development and to John Sninsky who, in addition to his contributions applying PCR to virus detection, was instrumental in the program to develop the *Taq* polymerase and automated thermocylers.   Finally, this book would not have been possible without the patience and dedication of Kathy Levenson who worked closely with the authors and, with the help of Dean Grantham, transformed the sporadic trickle of manuscripts into camera-ready copy.

Technology

Principles and Applications for DNA Amplification

Henry A. Erlich
EDITOR

New York   London   Tokyo   Melbourne   Hong Kong

For Brenda and Justin

Published in the United States and Canada by Stockton Press, 1989
15 East 26th Street, New York, N.Y. 10010

Library of Congress Cataloging-in-Publication Data

PCR technology : principles and applications for DNA amplification /
    Henry A. Erlich, editor.
        p.   cm. — (Breakthroughs in molecular biology)
    ISBN 0-935859-56-X (pbk.) : $19.95
    1. Polymerase chain reaction. 2. Gene amplification. I. Erlich.
Henry A., 1943-   . II. Series.
    [DNLM: 1. DNA Polymerases. 2. Gene Amplification. 3. Genetic
Engineering. 4. Molecular Biology. 5. RNA Polymerases. QH 442
P348]
QP606.D46P37 1989
574.87'3282 — dc 19
DNLM/DLC
for Library of Congress   88-4342
                                                               CIP

Published in the United Kingdom by
MACMILLAN PUBLISHERS LTD (Journals Division), 1989
Distributed by Globe Book Services Ltd,
Brunel Road, Houndmills, Basingstoke,
Hants RG21 2XS, England

British Library Cataloguing in Publication Data

Erlich, Henry
    DNA amplification using the PCR
    1. Organisms. DNA polymerases
    I. Title
    574.87'3282

    ISBN 0-333-48948-9

Printed in the United States of America
9 8 7 6 5 4

CONTRIBUTORS

James W. Ajioka, Ph.D.
Post-doctoral Fellow
Department of Genetics
Washington University School of Medicine

Norman Arnheim, Ph.D.
Professor of Molecular Biology
Department of Biological Sciences
University of Southern California

Edward T. Blake, D. Crim.
Forensic Science Associates

Johannes L. Bos, Ph.D.
Professor
Sylvius Laboratoria
University of Leiden, The Netherlands

Teodorica L. Bugawan
Research Associate
Department of Human Genetics
Cetus Corporation

C. Thomas Caskey, Ph.D.
Professor and Director, Institute for Molecular Genetics
Investigator, Howard Hughes Institute
Baylor College of Medicine

J.S. Chamberlain, Ph.D.
Post-doctoral Fellow
Institute for Molecular Genetics
Baylor College of Medicine

David R. Cox, M.D., Ph.D.
*Associate Professor of Psychiatry and
  Biochemistry/Biophysics
University of California, San Francisco*

B.L. Daugherty, M.S.
*Research Biochemist
Department of Cellular and Molecular Biology
Merck, Sharp and Dohme Research Laboratories*

Henry A. Erlich, Ph.D.
*Senior Scientist, Director
Department of Human Genetics
Cetus Corporation*

Dan Garza, Ph.D.
*Post-doctoral Fellow
Department of Genetics
Washington University School of Medicine*

David Gelfand, Ph.D.
*Vice-President, Scientific Affairs
Product Development, PCR Division
Cetus Corporation*

Richard A. Gibbs, Ph.D.
*Post-doctoral Fellow
Institute for Molecular Genetics
Baylor College of Medicine*

Ulf Gyllensten, Ph.D.
*Department of Medical Genetics
Biomedical Center, Uppsala, Sweden*

J. Han, Ph.D.
*Research Fellow
Department of Cellular and Molecular Biology
Merck, Sharp and Dohme Research Laboratories*

Daniel L. Hartl, Ph.D.
*McDonnell Professor of Genetics
Department of Genetics
Washington University School of Medicine*

Russell Higuchi, Ph.D.
*Scientist
Department of Human Genetics
Cetus Corporation*

Ernest Kawasaki, Ph.D.
*Scientist
Department of Human Genetics
Cetus Corporation*

Haig H. Kazazian, Jr., M.D.
Professor of Pediatrics, Gynecology, and Obstetrics
Genetics Unit, Department of Pediatrics
The Johns Hopkins Hospital

Thomas D. Kocher, Ph.D.
Post-doctoral Fellow
Department of Biochemistry
University of California, Berkeley

Shirley Kwok
Associate Scientist
Department of Infectious Diseases
Cetus Corporation

Simon W. Law, Ph.D.
Senior Research Fellow
Department of Cellular and Molecular Biology
Merck, Sharp and Dohme Research Laboratories

G.E. Mark, Ph.D.
Associate Director
Department of Cellular and Molecular Biology
Merck, Sharp and Dohme Research Laboratories

Richard M. Myers, Ph.D.
Assistant Professor
Department of Physiology and Biochemistry/Biophysics
University of California, San Francisco

David L. Nelson, Ph.D.
Post-doctoral Fellow
Howard Hughest Institute
Baylor College of Medicine

Howard Ochman, Ph.D.
Research Assistant Professor
Department of Genetics
Washington University School of Medicine

L. O'Neill, M.S.
Staff Biochemist
Department of Cellular and Molecular Biology
Merck, Sharp and Dohme Research Laboratories

Christian Oste, Ph.D.
Bioscience Specialist
Perkin-Elmer Cetus

Randall Saiki
Associate Scientist
Department of Human Genetics
Cetus Corporation

George Sensabaugh, Ph.D.
Professor and Chairman
Department of Public Health
University of California, Berkeley

Val C. Sheffield, M.D., Ph.D.
Post-doctoral Fellow
Department of Physiology and Biochemistry/Biophysics
University of California, San Francisco

John J. Sninsky, Ph.D.
Director
Department of Infectious Diseases
Cetus Corporation

J.-S. Tung, Ph.D.
Senior Research Fellow
Department of Cellular and Molecular Biology
Merck, Sharp and Dohme Research Laboratories

Cecilia H. von Beroldingen, Ph.D.
Post-doctoral Fellow
Department of Public Health
University of California, Berkeley

Alice Wang
Associate Scientist
Department of Molecular Biology
Cetus Corporation

Thomas J. White, Ph.D.
Senior Director of Diagnostics Research
Hoffmann La Roche Inc.

CONTENTS

ix

PART ONE

BASIC METHODOLOGY

In the short history of molecular biology, the emergence of a new technique (e.g., Southern blotting, molecular cloning, pulsed field gel electrophoresis) has often transformed the way we think about approaching both fundamental and applied biological problems. The capacity to amplify specific segments of DNA, made possible by the polymerase chain reaction (PCR), represents such a change. The PCR is an *in vitro* method for the enzymatic synthesis of specific DNA sequences, using two oligonucleotide primers that hybridize to opposite strands and flank the region of interest in the target DNA. A repetitive series of cycles involving template denaturation, primer annealing, and the extension of the annealed primers by DNA polymerase results in the exponential accumulation of a specific fragment whose termini are defined by the 5' ends of the primers. Because the primer extension products synthesized in one cycle can serve as a template in the next, the number of target DNA copies approximately doubles at every cycle. Thus, 20 cycles of PCR yields about a million-fold (2^{20}) amplification. This method, which was invented by Kary Mullis,[1,2] was originally applied by a group in the Human Genetics Department at Cetus to the amplification of

1

human β-globin DNA and to the prenatal diagnosis of sickle-cell anemia.[3,4,5]

Initially, the PCR used the Klenow fragment of *E. coli* DNA polymerase I to extend the annealed primers. This enzyme was inactivated by the high temperature required to separate the two DNA strands at the outset of each PCR cycle. Consequently, fresh enzyme had to be added during every cycle. The introduction of the thermostable DNA polymerase (*Taq* polymerase) isolated from *Thermus aquaticus* (see Chapter 2) transformed the PCR into a simple and robust reaction which could now be automated by a thermal cycling device (see Chapter 3). The reaction components (template, primers, *Taq* polymerase, dNTP's, and buffer) could all be assembled and the amplification reaction carried out by simply cycling the temperature within the reaction tube.[6] The effect of varying the reaction parameters (e.g., enzyme, primer and Mg^{++} concentration as well as the temperature cycling profile) on the specificity and yield of the amplification is discussed in Chapter 1. Although, for any given pair of oligonucleotide primers, an optimal set of conditions can be established, there is no single set of conditions that will be optimal for all possible reactions.

The specificity of the PCR is typically analyzed by evaluating the production of the target fragment relative to other products by gel electrophoresis. Another factor influencing the homogeneity of the PCR product is the *concentration* of the target sequence in the genomic template.[6] This effect was revealed by experiments[6] in which normal genomic DNA (with two copies of the β-globin gene/cell) and genomic DNA with a homozygous deletion of β-globin were amplified with β-globin primers. Using the normal genomic template, this reaction generated a unique β-globin fragment with no detectable nonspecific products. Using the β-globin deletion DNA sample, no β-globin fragment was synthesized, as expected; however, several non-target fragments were produced. Thus, in the absence of the "right" template, the "wrong" sequences became amplified by the β-globin primers, illustrating the old adage, "Idle hands are the Devil's playthings." This effect accounts, in part, for the generally heterogenous gel profile of PCR products synthesized from a very rare template sequence, like HIV sequences which are present in only a small fraction of the cells sampled. As discussed in Chapter 1, modification of various PCR parameters to optimize the specificity of amplification may yield more homogenous products even in rare template reactions.

The initial PCR method based on DNA synthesis by the Klenow enzyme at 37°C was not highly specific. Although a specific target fragment could be amplified up to a million-fold, most of what was synthesized in the PCR was not, in fact, this target fragment. By cloning a β-globin amplification reaction and screening the individual clones with a β-globin probe to detect the target sequence and with one of primers to detect *any* amplified sequence, the specificity of the PCR was estimated to be ~1%.[7] Other primer pairs were somewhat more or less specific but, in general, the Klenow enzyme PCR was not a very specific reaction and required subsequent analysis with a specific hybridization probe[3,4] or, in some cases, with internal "nested primers"[1] to detect and characterize the amplified target sequence.

The use of the *Taq* polymerase not only simplified the PCR procedure (see above) but significantly increased the specificity and the overall yield of the reaction. The higher temperature optimum for the *Taq* polymerase (~75°C) allowed the use of higher temperatures for primer annealing and extension, thereby increasing the overall stringency of the reaction and minimizing the extension of primers that were mismatched with the template. At 37°C, many of these mismatched primers are sufficiently stable to be extended by the Klenow enzyme, resulting in non-specific amplification products. The increase in the specificity of the *Taq* PCR results in an improved yield of the amplified target fragment by reducing the competition by non-target products for enzyme and primers. In the later cycles, the amount of enzyme is no longer sufficient to extend all the annealed primer/template complexes in a single cycle period, resulting in a reduced efficiency and a "plateau" in the amplification reaction. This plateau is reached later (e.g., about 30 cycles rather than 20 starting with 1 μg of genomic DNA) in the *Taq* PCR than in the reaction with the Klenow enzyme due to the increased specificity of the former reactions. Other factors like the reassociation of the template strands at high product concentration may also contribute to the plateau effect and are discussed in Chapter 1. In addition to the increase in the specificity and yield of PCR made possible by *Taq* polymerase, the use of this enzyme allows the amplification of much longer fragments (up to 10 kb, albeit with reduced efficiency)[8] than does the Klenow enzyme (<400 bp).

Although the PCR is considered primarily a method for producing copies of a specific sequence, it is also a very powerful and precise way of altering a particular template sequence. Since the oligonucleotide primers become physically incorporated into the amplified product and mismatches between the 5' end of the primer and *initial* template[†] are tolerated, it is possible to introduce new sequence information adjacent to the target sequence via the primers. Thus, for cloning a given sequence, one is no longer constrained by the restriction sites Nature provides but can add any restriction enzyme recognition sequence to the 5' ends of the primer[7] leading to the formation of a new restriction site in the double-stranded amplification product. Similarly, regulatory elements such as a T7 promoter can be added,[1] allowing the synthesis of RNA copies from the PCR product using T7 RNA polymerase. Furthermore, specific nucleotide substitutions, insertions, and deletions can also be introduced into the amplified product with the appropriate primers.

Unlike these directed mutations, changes introduced into the sequence of the PCR products due to nucleotide misincorporation can create potential problems. An estimate of the fidelity of *Taq* polymerase during the PCR was made by the sequence analysis of multiple M13 clones generated by cloning the products of a particular amplification reaction. In this initial study,[6] the *frequency* of errors was found to be ~1/400 and the error *rate*

[†]After the first few cycles, virtually all of the templates have been synthesized in previous cycles and, therefore, contain the primer sequences.

was calculated to be ~2×10^{-4} nt/cycle. More recent studies, carried out on a different gene using the current, optimized protocol (reduced Mg^{++} and dNTP concentrations), found a lower frequency of errors.[9] The original estimate is in reasonable agreement with the report of a 10^{-4} nt error rate determined by the *in vitro* replication of a β-galactosidase template by the *Taq* polymerase.[10] Although different template sequences may have a somewhat different "mutability" and different reaction conditions may influence the fidelity of the *Taq* polymerase, the original "high" error rate estimated for *Taq* polymerase PCR (~10^{-4}) does not pose a problem for most applications. In the analysis of the *population* of amplified products, as in oligonucleotide probe hybridization or in direct sequencing, the rare errors in individual products are not detectable. However, in the sequence analysis of individual clones derived from a PCR, sequences must be determined from multiple clones to distinguish misincorporated nucleotides from the faithful copes of the template sequence. Other DNA polymerases with an active "proof-reading" function, like T4 DNA polymerase, can carry out PCR[11] and may prove useful in studies where a very low error rate is required. As discussed in Chapter 2, the *Taq* polymerase does not contain measurable 3'-to-5' exonuclease "proof-reading" activity.

An important property of the PCR, particularly in diagnostic applications, is the capacity to amplify a target sequence from crude DNA preparations[4] as well as from degraded DNA templates.[12] The DNA in a sample need not be chemically pure to serve as a template provided that the sample does not contain inhibitors of *Taq* polymerase. The effect of various rapid protocols for sample preparation on the PCR is discussed in Chapter 4. The ability to amplify specific sequences from crude DNA samples has important implications for research applications, (e.g., sperm lysates[13]; see Chapter 12), for medical diagnostic applications (e.g., mouthwash[14] or archival paraffin-embedded tissue samples[15]) and for forensics (e.g., individual hairs[16]; see Chapter 17).

The possibility of contamination in the amplification reaction is an issue with broad implications for both research and diagnostic applications. Given the capacity of PCR to synthesize millions of DNA copies, contamination of the sample reaction with either products of a previous reaction (product carryover) or with material from an exogenous source is a potential problem - particularly in those reactions initiated with only a few templates. The amplification of individual sperm (see Chapter 12), of single hairs (see Chapter 17), and of HIV genomic sequences (see Chapter 19) all require rigorous measures to minimize and monitor potential contamination. In general, careful laboratory procedure, prealiquoting reagents, the use of positive displacement pipettes, and the physical separation of the reaction preparation from the analysis of the reaction products are all precautions that reduce the risk. Carrying out only the minimal number of PCR cycles required for analysis also minimizes the chance that a rare contaminating template will be amplified. A panel of "blank" reactions with no template DNA is necessary to detect potential contamination. In genetic typing, a sample that has been contaminated can often be identified by a genotyping result with more than 2 alleles (see Chapters 16 and 17).

The PCR, like recombinant DNA technology, has had an enormous impact in both basic and diagnostic aspects of molecular biology because it can produce large amounts of a specific DNA fragment from small amounts of a complex template. Recombinant DNA techniques create molecular clones by conferring on a specific sequence the ability to replicate by inserting it into a vector and introducing the vector into a host cell. PCR represents a form of "*in vitro* cloning" that can generate, as well as modify, DNA fragments of defined length and sequence in a simple automated reaction.

In its early, heady days, Erwin Chargaff once described molecular biology dismissively as "the practice of biochemistry without a license." Its spectacular success over the last 30 years suggests that significant progress can be made even in the absence of an established theoretical foundation. One element in the rapid development of molecular biology has been its characteristic embrace of new techniques and the aggressive pursuit of the new questions made possible by their use. PCR has facilitated molecular analysis in many disparate fields of biological research; given the ease and simplicity of PCR amplification, it can be said to allow the practice of molecular biology without a permit. It is likely that future practitioners will continue to find new applications and modifications of this powerful technology.

REFERENCES

1. Mullis, K.B., and Faloona, F. (1987) *Meth. Enzymol.* 155:335.
2. Mullis, K.B., Faloona, F., Scharf, S.J., Saiki, R.K., Horn, G.T., and Erlich, H.A. (1986) *Cold Spring Harbor Symp. Quant. Biol.* 51:263-273.
3. Saiki, R., Scharf, S., Faloona, F., Mullis, K., Horn, G., Erlich, H.A., and Arnheim, N. (1985) *Science* 230:1350.
4. Saiki, R.K., Bugawan, T.L., Horn, G.T., Mullis, K.B., and Erlich, H.A. (1986) *Nature* 324:163.
5. Embury, S.H., Scharf, S.J., Saiki, R.K., Gholson, M.A., Golbus, M., Arnheim, N., and Erlich, H.A. (1987) *New Engl. J. Med.* 316:656.
6. Saiki, R.K., Gelfand, D.H., Stoffel, S., Scharf, S., Higuchi, R.H., Horn, G.T., Mullis, K.B., and Erlich, H.A. (1988) *Science* 239:487.
7. Scharf, S.J., Horn, G.T., and Erlich, H.A. (1986) *Science* 223:1076.
8. Jeffreys, A.J., Wilson, V., Neumann, R., and Keyte, J. (1988) *Nucl. Acids Res.* 16:10953-10971.
9. Goodenow, M., Huet, T., Saurin, W., Kwok, S., Sninsky, J., and Wain-Hobson, S. (1989) *J. Acquired Immune Deficiency Syndromes,* in press.
10. Tindall, K.R., and Kunkel, T.A. (1988) *Biochem.* 27:6008-6013.
11. Keoharong, P., Kat, A., Cariello, N.F., and Thilly, W.G. (1988) *DNA* 7:63-70.
12. Bugawan, T.L., Saiki, R.K., Levenson, C.H., Watson, R.M., and Erlich, H.A. (1988) *Bio/Technology* 6:943.
13. Li, H., Gyllensten, U.B., Cui, X., Saiki, R.K., Erlich, H.A., and Arnheim, N. (1988) *Nature* 335:414.
14. Lench, N., Stanier, P., and Williamson, R. (1987) *Lancet* ii:1356-1358.
15. Shibata, D.K., Martin, J.W., and Arnheim, N. (1988) *Cancer Res.* 48:4564-4566.
16. Higuchi, R., von Beroldingen, C.H., Sensabaugh, G.F., and Erlich, H.A. (1988) *Nature* 332:543

4- Phenol burn your hand, also denatures sample turn DNA off your hand.

3- Disrupts Protein Structures by forming H bonds by breaking open protein coat.

CHAPTER 1

The Design and Optimization of the PCR

Randall K. Saiki

2 - Solubilies the lipid membrane causing a disruption of Protein Structure which separate + can then Centrifuged.

INTRODUCTION

In the few years since its introduction,[1,2,3] the polymerase chain reaction has already become a widespread research technique. Like the PCR itself, the numbers of its practitioners have been accumulating exponentially and will probably continue to do so in the near future as the method finds wider applications in fields other than molecular biology. This popularity of the PCR is primarily due to its apparent simplicity and high probability of success. Reduced to its most basic terms, PCR merely involves combining a DNA sample with oligonucleotide primers, deoxynucleotide triphosphates, and the thermostable *Taq* DNA polymerase in a suitable buffer, then repetitively heating and cooling the mixture for several hours until the desired amount of amplification is achieved.

In fact, the PCR is a relatively complicated and, as yet, incompletely understood biochemical brew where constantly changing kinetic interactions among the several components determine the quality of the products obtained. Although the results will be good in most cases, there are a

7

number of parameters that can be explored if better results are required or if the reaction fails altogether. This chapter will examine some of those parameters and include some guidelines for the design and execution of the PCR.

THE "STANDARD" REACTION

Because of the wide variety of applications in which PCR is being used, it is probably impossible to describe a single set of conditions that will guarantee success in all situations. Nevertheless, the reaction outlined below should prove to be adequate for most amplifications and for those cases where it isn't, it at least defines a common starting point from which changes can be attempted.

The standard PCR is typically done in a 50- or 100-µl volume and, in addition to the sample DNA, contains 50 mM KCl, 10 mM Tris.HCl (pH 8.4 at room temp.), 1.5 mM $MgCl_2$, 100 µg/ml gelatin, 0.25 µM of each primer, 200 µM of each deoxynucleotide triphosphate (dATP, dCTP, dGTP, and dTTP), and 2.5 units of *Taq* polymerase. The type of the DNA sample will be variable, of course, but it will usually have between 10^2 to 10^5 copies of template (e.g., 0.1 µg human genomic DNA). A few drops of mineral oil are often added to seal the reaction and prevent condensation.

The amplification can be conveniently performed in a DNA Thermal Cycler (Perkin-Elmer Cetus Instruments) using the "Step-Cycle" program set to denature at 94°C for 20 sec, anneal at 55°C for 20 sec, and extend at 72°C for 30 sec for a total of 30 cycles. (The "Step-Cycle" program causes the instrument to heat and cool to the target temperatures as quickly as possible. In the current instrument, this results in a heating rate of about 0.3°C per sec and a cooling rate of about 1°C per sec, for an overall single cycle time of approximately 3.75 min.)

These conditions can be used to amplify a wide range of target sequences with excellent specificity (Figure 1). However, for those reactions where the conditions described above do not produce the desired results, the following sections describe some ways in which a PCR can be improved.

PRIMER SELECTION

Unfortunately, the approach to the selection of efficient and specific primers remains somewhat empirical. There is no set of rules that will ensure the synthesis of an effective primer pair. Yet it is the primers more than anything else that determine the success or failure of an amplification reaction. Fortunately, the majority of primers can be made to work and the following guidelines will help in their design.

1. Where possible, select primers with a random base distribution and with a GC content similar to that of the fragment being amplified. Try to avoid primers with stretches of polypurines, polypyrimidines, or other unusual sequences.

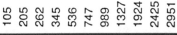

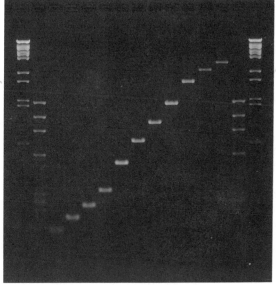

Figure 1. Amplification of β-globin fragments ranging from 150 to 2951 bp. Samples of 100 μl containing standard buffer, 200 μM each dNTP, and 250 nM each primer, 100 ng human genomic DNA, and 2.5 units *Taq* polymerase (Perkin Elmer/Cetus) were subjected to 30 cycles of amplification. Two μl of each sample were resolved on a 1.6% agarose gel and visualized by ethidium bromide fluorescence. The lengths of the products are indicated at the top of each lane. The markers are *Bst*EII-digested phage λ (500 ng) and *Hae*III-digested φx174-RF (250 ng).

2. Avoid sequences with significant secondary structure, particularly at the 3'-end of the primer. Computer programs, such as Squiggles or Circles available from the University of Wisconsin, are very useful for revealing these structures.

3. Check the primers against each other for complementarity. In particular, avoiding primers with 3' overlaps will reduce the incidence of "primer dimer" (see below).

Sometimes there will be constraints on where the primers can be located. For example, perhaps only a limited amount of sequence information is available. Under those circumstances, it is worthwhile to go ahead and try the primers. These guidelines only increase the chances that any given pair of oligonucleotides function properly, they are not absolute requirements.

Most primers will be between 20 and 30 bases in length and the optimal amount to use in an amplification will vary. Longer primers may be synthesized but are seldom necessary. Sequences not complementary to the template can be added to the 5'-end of the primers. These exogenous sequences become incorporated into the double-stranded PCR product and provide a means of introducing restriction sites[4] or regulatory elements (e.g., promoters) at the ends of the amplified target sequence.[5] If required, shorter primers or degenerate primers can be used as long as the thermal profile of the reaction is adapted to reflect the lower stability of the primed template. (For highly degenerate primers, it is preferable that the most unambiguous sequence be situated at the 3'-end of the primer, even to the extent of synthesizing a multiple series in which the various permutations of the 3' sequence are held constant.) In general, concentrations ranging from 0.05 to 0.5 µM of each oligonucleotide should be acceptable.

"Primer dimer" is an amplification artifact often observed in the PCR product, especially when many cycles of amplification are performed on a sample containing very few initial copies of template. It is a double-stranded fragment whose length is very close to the sum of the two primers and appears to occur when one primer is extended by the polymerase over the other primer. The resulting concatenation is an extremely efficient PCR template that can, if it occurs at an early cycle, easily overwhelm a reaction and become the predominant product.

The exact mechanism by which primer dimer is formed is not completely clear. The observation that primers with complementary 3'-ends are predisposed to dimer formation suggests that transient interactions that bring the termini in close proximity are the initiating event.[6] Several polymerases, including *Taq*, have been shown to have a weak non-template directed polymerization activity which can attach additional bases to a blunt-ended duplex.[7,8] If such an activity were also to occur on a single-stranded oligonucleotide, there is a good chance that the extension would form a short 3' overlap with the other primer sufficient to promote dimerization. In any event, if dimers present an obstacle, they can be reduced somewhat by using minimal concentrations of primers and enzyme.

Finally, although most primers will work with varying degrees of success, occasional primers will be synthesized that completely fail to amplify their intended target. The reasons for this remain somewhat obscure, but in many of these instances, simply moving the primers by a few bases in either direction will solve the problem.

THE PCR BUFFER

Changes to the PCR reaction buffer will usually effect the outcome of the amplification. In particular, the concentration of $MgCl_2$ can have a profound effect on the specificity and yield of an amplification.

Concentrations of about 1.5 mM are usually optimal (with 200 μM each dNTP), but in some circumstances, different amounts of Mg^{2+} may prove to be necessary (Figure 2). Generally, excess Mg^{2+} will result in the accumulation of non-specific amplification products and insufficient Mg^{2+} will reduce the yield. More recently, it has been shown that the reduction or elimination of KCl and gelatin can be beneficial.[9,10] Some protocols include 10% dimethyl sulfoxide (DMSO) ostensibly to reduce the secondary structure of the target DNA; our own experience has shown that DMSO can be slightly inhibitory to *Taq* polymerase and decrease the overall yield of amplification product.

The deoxynucleotide triphosphates (dATP, dCTP, dGTP, and dTTP) are usually present at 50 to 200 μM of each. Higher concentrations may tend to promote misincorporations by the polymerase (i.e., "thermodynamic infidelity") and should be avoided.[11] At 50 and 200 μM, there is sufficient precursor to synthesize approximately 6.5 and 25 μg of DNA, respectively.

Neutralized dNTP solutions can be obtained from several sources (e.g., USB, Sigma, Pharmacia). Considerable savings can be realized by purchasing lyophilized powders and preparing your own aqueous solutions, but they must be neutralized with sodium hydroxide and concentrations accurately determined by UV absorbance before use.

As deoxynucleotide triphosphates appear to quantitatively bind Mg^{2+}, the amount of dNTPs present in a reaction will determine the amount of free magnesium available. In the standard reaction, all four triphosphates are added to a final concentration of 0.8 mM; this leaves 0.7 mM of the original 1.5 mM $MgCl_2$ not complexed with dNTP. Consequently, if the dNTP concentration is changed significantly, a compensatory change in $MgCl_2$ may be necessary.

Taq polymerase is available from a number of vendors (e.g., Perkin-Elmer Cetus Instruments, New England Biolabs, Stratagene). The concentration of enzyme typically used in PCR is about 2.5 units per 100 μl reaction. For amplification reactions involving DNA samples with high sequence complexity, such as genomic DNA, there is an optimum concentration of *Taq* polymerase, usually 1 to 4 units per 100 μl. Increasing the amount of enzyme beyond this level can result in greater production of non-specific PCR products and reduced yield of the desired target fragment (Figure 3).

CYCLING PARAMETERS

PCR is performed by incubating the samples at three temperatures corresponding to the three steps in a cycle of amplification-denaturation, annealing, and extension. This cycling can be accomplished either manually with pre-set water baths, or automatically with the DNA Thermal Cycler.

In a typical reaction, the double-stranded DNA is denatured by briefly heating the sample to 90-95°C, the primers are allowed to anneal to their complementary sequences by briefly cooling to 40-60°C, followed by heating to 70-75°C to extend the annealed primers with the *Taq* polymerase.

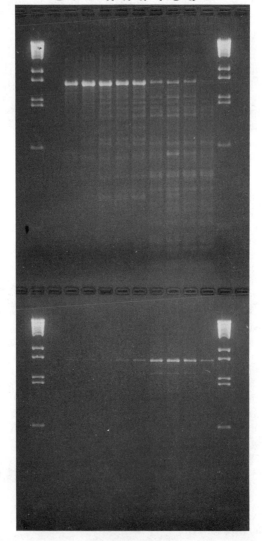

Figure 2. Effect of Mg^{2+} concentration on PCR specificity and yield. Two overlapping primer pairs that amplify 1.8-kb fragments of the human β-globin gene were titrated with various concentrations of MgCl$_2$. Amplifications were performed as described in Figure 1 except that MgCl$_2$ was varied from 0.5 to 10 mM (as indicated at the top of each lane). Although similar in size and from the same region of the β-globin gene, the two primer pairs have very different magnesium optima.

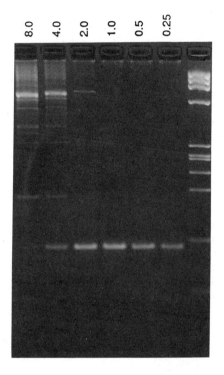

Figure 3. Effect of *Taq* polymerase concentration on PCR specificity and yield. Amplification of a 110-bp fragment of the β-globin gene was performed as described in Figure 1 but with levels of *Taq* DNA polymerase ranging from 0.25 to 8 units per 100-µl reaction. The optimum concentration of polymerase is in the neighborhood of 1 to 2 units.

The time of incubation at 70-75°C varies according to the length of target being amplified; allowing 1 min for each kilobase of sequence is almost certainly excessive, but it is a good place to begin. Shorter times should be tried once the other amplification conditions have been established. (The extension step can be eliminated altogether if the target sequence is approximately 150 bases or less. The polymerase retains significant activity at lower temperatures[9] and complete extension will occur during the thermal transition from annealing to denaturation.)

The ramp time, or time taken to change from one temperature to another, depends on the type of equipment used. With some notable exceptions, this rate of temperature change is not important and the fastest ramps attainable are used to shorten the cycle time. However, in order to be certain that the samples reach the intended temperatures, the actual ramp times for a particular setup should be determined by measuring the *sample* temperature during a test amplification. A microprobe thermocouple (Cole-Parmer) and digital multimeter are very helpful for this purpose.

Insufficient heating during the denaturation step is a common cause of failure in a PCR reaction. It is very important that the reaction reaches a temperature at which complete strand separation occurs. A temperature of about 94°C should be adequate in most cases. As soon as the sample reaches 94°C, it can be cooled to the annealing temperature. Extensive denaturation is probably unnecessary and limited exposure to elevated temperatures helps maintain maximum polymerase activity throughout the reaction.

The temperature at which annealing is done depends on the length and GC content of the primers. A temperature of 55°C is a good starting point for typical 20-base oligonucleotide primers with about 50% GC content; even higher temperatures may be necessary to increase primer specificity. Because of the very large molar excess of primers present in the reaction mix, hybridization occurs almost instantaneously and long incubation at the annealing temperature is not required. (The 20 sec denaturation and annealing incubation times used with the Step-Cycle program on the Thermal Cycler is the amount of time it takes for a 100-μl reaction in a 0.5-ml microcentrifuge tube to equilibrate with the block temperature.)

In some cases, primers of only 12 to 15 bases are available and an annealing temperature around 40-45°C is needed. However, primers of that length are unlikely to remain annealed at the 72°C extension temperature. The problem can be overcome by taking advantage of the partial enzymatic activity of the polymerase at lower temperatures to extend the primers by several bases and stabilize them. This is accomplished either by an intermediate incubation at 50-60°C or by heating gradually from 40°C to 72°C. Degenerate primers will often have multiple mismatches with their target sequence and should be treated in a similar manner.

It is often possible to anneal and extend the primers at the same temperature.[12] In addition to simplifying the procedure to a two-temperature cycle, simultaneously annealing and extending at a temperature greater than 55°C may further improve the specificity of the reaction.

AMPLIFICATION PLATEAU

The amplification reaction is not infinite. After a certain number of cycles the desired amplification fragment gradually stops accumulating exponentially and enters a linear or stationary phase. This second stage of the reaction is called the "plateau." The point at which a PCR reaction reaches its plateau depends primarily on the number of copies of target originally present in the sample and by the total amount of DNA synthesized. (As such, it is not completely informative to specify the performance of a reaction solely by degree of amplification without also indicating the initial template concentration.)

In addition to the mundane possibilities such as exhaustion of primer or dNTP or inactivation of polymerase or dNTP, (none of which is significant in a standard reaction), there remains three more exotic causes of plateau -

substrate excess conditions, competition by non-specific products, and product reassociation.

Substrate excess is simply the result of having synthesized more DNA than the amount of *Taq* polymerase present in the reaction is capable of replicating in the allotted extension time. For a standard 100-μl reaction containing 2.5 units of *Taq* polymerase, substrate excess conditions begin to occur around 1 μg of DNA (3 nmol of deoxynucleotide monophosphate). It can be overcome by increasing the extension time and/or increasing the amount of enzyme in the reaction. This is usually not practical, however, because each succeeding cycle would require the doubling of extension time and/or polymerase to continue exponential growth.

Competition by non-specific amplification products is closely related to substrate excess conditions. In this case, the unwanted DNA fragments compete with the desired fragment for the attention of the limiting DNA polymerase. Clearly, this problem can be alleviated by increasing the specificity of the reaction so that the non-target sequences are not allowed to accumulate to any significant degree.

Alternatively, further accumulation of product can be attenuated by reassociation of the single-stranded PCR fragments before the annealed primers can be extended. Whether this inhibition is due to branch migration and displacement of the primer or by inefficient displacement synthesis on the part of the polymerase itself is unclear. This limitation usually comes into play when the product concentration approaches 10 pmol per 100 μl and is difficult to avoid except by dilution of the reaction.

In most of these cases, the plateau is an unavoidable and inherent limitation of the PCR reaction. Happily, by the time it occurs sufficient amounts of product will have accumulated for almost any purpose. For those few situations where even more material is needed, it is probably much easier to set up multiple reactions than to try to evade the plateau.

FACTORS AFFECTING SPECIFICITY

There are a number of factors that can affect the specificity of the amplification reaction. The stringency of the annealing step can be controlled to some extent by adjusting the annealing temperature. Minimizing the incubation time during the annealing and extension steps will limit the opportunities for mispriming and extension by molecules of otherwise idle DNA polymerase. Reducing primer and enzyme concentrations also serves to limit mispriming, particularly the type that leads to dimerization. Finally, changing $MgCl_2$ (and perhaps KCl) levels can further improve specificity, either by increasing the stringency of the reaction or by direct effects on the polymerase itself (activity, processivity, etc.).

SUMMARY

Because of the complex interactions among the components of a PCR, (particularly between the primers and the DNA sample), and the wide variety of applications in which this technique is being used, it is highly unlikely that there ever will be one set of amplification conditions that will prove optimal for all situations. In most cases, however, the general conditions described here have usually provided us with satisfactory results. When necessary, minor adjustments to these parameters will often transform a marginal PCR into one with excellent specificity and yield.

ACKNOWLEDGMENTS

The author would like to thank David Gelfand for valuable discussions about PCR.

REFERENCES

1. Saiki, R.K., Scharf, S., Faloona, F., Mullis, K.B., Horn, G.T., Erlich, H.A., and Arnheim, N. (1985) *Science* 37:170-172.
2. Mullis, K.B., Faloona, F.A., Scharf, S.J., Saiki, R.K., Horn, G.T., and Erlich, H.A. (1986) *Cold Spring Harbor Symp. Quant. Biol.* 51:263-273.
3. Faloona, F., and Mullis, K. (1987) *Meth. Enzymol.*155:335-350.
4. Scharf, S.J., Horn, G.T., and Erlich, H.A. (1986) *Science* 233:1076-1078.
5. Stoflet, E.S., Koeberl, D.D., Sarkar, G., and Sommer, S.S. (1988) *Science* 239:491-494.
6. Watson, B., personal communication.
7. Clark, J.M. (1988) *Nucl. Acids Res.* 16:9677-9686.
8. Denney, D., personal communication.
9. Innis, M.A., Myambo, K.B., Gelfand, D.H., and Brow, M.A. (1988) *Proc. Natl. Acad. Sci. USA* 85:9436-9440.
10. McCabe, P., personal communication.
11. Petruska, J., Goodman, M.F., Boosalis, M.S., Sowers, L.C., Cheong, C., and Tinoco, I. (1988) *Proc. Natl. Acad. Sci. USA* 85:6252-6256.
12. Kim, H.S., and Smithies, O. (1988) *Nucl. Acids Res.* 16:8887-8903.

6a ☐ 75 — 25% Pure DNA?   c   ☐ ← Concatomeric
b) ☐   —. Supercoiled
   ○   ≋ ll RNA
   ☐   ≋ — mRNA
tRNA → ⌡≋ RNA (tRNA, rRNA, mRNA).   ⌐ rRNA
       ⌡≋ mRNA - Flourescent Band.

CHAPTER 2

Taq DNA Polymerase

David H. Gelfand

The polymerase chain reaction (PCR) method for amplifying selectively discrete segments of DNA has found wide-spread applications in molecular biology, due in part, to the substitution of a thermostable DNA polymerase isolated from *Thermus aquaticus (Taq)*[1] for the previously used *E. coli* DNA Polymerase I, Klenow fragment (PolI-Kf).[2,3] Since *Taq* DNA Polymerase can withstand repeated exposure to the high temperatures (94°-95°C)[1] required for strand separation, the tedium and frequent rebellion resulting from having to add PolI-Kf after each cycle is minimized.

ENZYMATIC CHARACTERISTICS AND PROPERTIES

T. aquaticus strain YT1, a thermophilic, eubacterial microorganism capable of growth at 70°-75°C., was isolated from a hot spring in Yellowstone National Park and first described twenty years ago.[4] DNA polymerase activities with an estimated molecular mass of 60-68 kDa and an inferred specific activity of 2,000-8,000 units/mg have been isolated previously from this organism.[5,6] In contrast, we have isolated a DNA polymerase activity from *T. aquaticus* with a specific activity of 200,000 U/mg (S. Stoffel, in preparation) which migrates on SDS-PAGE slightly faster than Phosphorylase B (97.3 kDa) and which has an inferred molecular weight (based on DNA sequence information) of 93,910.[7]

As observed for several DNA polymerase activities isolated from thermophilic microorganisms, 94 kDa *Taq* DNA Polymerase has a relatively high temperature optimum (T_{opt}) for DNA synthesis. Depending on the nature of the DNA template, we have found an apparent T_{opt} of 75-80°C with a K_{cat} approaching 150 nucleotides/sec/enzyme molecule. Innis *et al.*[1] have reported highly processive synthesis properties and an extension rate of >60 nt/sec at 70°C with *Taq* DNA Polymerase for a GC-rich 30-mer primer on M13 and significant extension activity at 55°C (24 nt/sec). Even at lower temperatures, *Taq* DNA Polymerase has extension activities of ~0.25 and 1.5 nt/sec at 22° and 37°C, respectively. At lower temperatures, there is a marked attenuation in the apparent processivity of *Taq* DNA polymerase. This could reflect an impaired ability of *Taq* polymerase to extend through regions of local intramolecular secondary structure on the template strand or a change in the ratio of the forward rate constant to the dissociation constant. Very little DNA synthesis is seen at very high temperatures (>90°C). DNA synthesis at higher temperatures *in vitro* may be limited by the stability of the primer or priming-strand and the template-strand duplex.

Although *Taq* DNA polymerase has a very limited ability to synthesize DNA above 90°C, the enzyme is relatively stable to and is not denatured irreversibly by exposure to high temperature. In a PCR mix, *Taq* DNA Polymerase retains about 50% of its activity after 130 min, 40 min, and 5-6 min at 92.5°, 95°, 97.5°C, respectively (B. Watson, personal communication). Preliminary results indicate retention of 65% activity after a 50-cycle PCR when the upper limit temperature (in tube) is 95°C for 20 sec in each cycle.

Taq DNA polymerase activity is sensitive to the concentration of magnesium ion as well as to the nature and concentration of monovalent ions (DHG, in preparation). Using minimally activated salmon sperm DNA as template in a standard 10 min assay,[7] 2.0 mM magnesium chloride maximally stimulates *Taq* polymerase activity at 0.7-0.8 mM total dNTP. Higher concentrations of Mg^{++} are inhibitory, with 40-50% inhibition at 10 mM $MgCl_2$. Since deoxynucleotide triphosphates can bind Mg^{++}, the precise magnesium concentration that is required to maximally activate the enzyme is dependent on the dNTP concentration. In addition, at the correspondingly optimal magnesium concentration, the synthesis rate of *Taq* polymerase decreases by as much as 20-30% as the total dNTP concentration is increased to 4-6 mM. This observation may result from substrate inhibition.

Low, balanced concentrations of dNTPs have been observed to give satisfactory yields of PCR product, to result frequently in improved specificity, to facilitate labeling of PCR products with radioactive or biotinylated precursors, and to contribute to increased fidelity of *Taq* polymerase (see below). In a 100-μl PCR with 40 μM each dNTP, there are sufficient nucleotide triphosphates to yield 2.6 μg of DNA when only half of the available dNTPs are incorporated into DNA. It is likely that very low dNTP concentrations may adversely affect the processivity of *Taq* DNA polymerase. Furthermore, the precise concentration of free and enzyme-bound magnesium may affect the processivity of *Taq* polymerase as has been inferred for calf thymus DNA polymerases α and δ.[9]

Purified 94 kDa *Taq* DNA polymerase does not contain an inherent 3'-5'-exonuclease activity (S. Stoffel).[10] Single nucleotide incorporation/misincorporation, biochemical fidelity measurements have indicated that the ability of "non-proofreading" DNA polymerases to misincorporate a deoxynucleotide triphosphate is determined critically by the concentration of that triphosphate.[11] Accordingly, a model of "Km or Vmax discrimination" has been advanced to suggest mechanisms by which non-proofreading DNA polymerases may achieve high fidelity. Similar data have been obtained with regard to extension of a mismatched primer/template.[12] Although not yet measured kinetically, *Taq* DNA polymerase appears to extend a mismatched primer/template significantly less efficiently than a correct primer/template (Figure 2 in ref. 8). It is not known if *T. aquaticus* contains a separate 3'-5'-exonuclease activity which may be associated with the polymerase *in vivo*. Since the purification protocol for native *Taq* DNA polymerase (DHG, in preparation) was intended to yield a single polypeptide chain enzyme, we could have failed to detect an *E. coli* Pol III "ε-like" associated subunit.

Taq DNA polymerase has a DNA synthesis-dependent, strand replacement, 5'-3'- exonuclease activity. There is little, if any, degradation of a 5' ^{32}P-labeled oligodeoxynucleotide, either as single-stranded DNA or where annealed to an M13 template. Furthermore, the presence of a "blocking," annealed, non-extendable oligodeoxynucleotide "primer" (3'-phosphorylated during synthesis) fails to attenuate incorporation from a 3'-OH terminated upstream primer. There is little, if any, displaced "blocking primer," and the products of exonuclease action are primarily deoxynucleoside monophosphate (85%) and dinucleoside phosphate (15%, S. Stoffel, unpublished).

Modest concentrations of KCl stimulate the synthesis rate of *Taq* polymerase by 50-60% with an apparent optimum at 50 mM. Higher KCl concentrations begin to inhibit activity, and no significant activity is observed in a DNA sequencing reaction at ≥75 mM KCl[8] or in a 10-min incorporation assay at >200 mM KCl.

The addition of either 50 mM ammonium chloride or ammonium acetate or sodium chloride to a *Taq* DNA polymerase activity assay results in mild inhibition, no effect, or slight stimulation (25-30%), respectively.

Low concentrations of urea, DMSO, DMF, or formamide have no effect on *Taq* polymerase's incorporation activity (Table 1). The presence of 10% DMSO (used previously in Klenow-mediated PCR)[13] in a 70°C *Taq* Polymerase activity assay inhibits DNA synthesis by 50%. While several investigators have observed that inclusion of 10% DMSO facilitates certain PCR assays, it is not clear which parameters of PCR are affected. The presence of DMSO may affect the Tm of the primers, the thermal activity

Table 1. Inhibitor Effects on *Taq* Pol I Activity

Inhibitor	Concentration	Activity*
Ethanol	≤3%	100%
	10%	110%
Urea	≤0.5 M	100%
	1.0 M	118%
	1.5 M	107%
	2.0 M	82%
DMSO	≤1%	100%
	10%	53%
	20%	11%
DMF	≤ 5%	100%
	10%	82%
	20%	17%
Formamide	≤10%	100%
	15%	86%
	20%	39%
SDS	.001%	105%
	.01%	10%
	.1%	<.1%

*dNTP incorp. activity at 70° with Salmon Sperm DNA/10 min

profile of *Taq* DNA Polymerase and/or the degree of product strand separation achieved at a particular "denaturation" or upper-limit temperature. Curiously, 10% ethanol fails to inhibit *Taq* activity and 1.0 M urea stimulates *Taq* activity. These effects on incorporation activity may not reflect the degree to which these agents affect the PCR. For example urea at 0.5 M completely inhibits a PCR assay (C.-A. Chang, personal communication). Finally, the inhibitory effects of low concentrations of SDS can be completely reversed by high concentrations of certain nonionic detergents (e.g., Tween 20 and NP40). Thus 0.5% each Tween 20/NP40 instantaneously reverses the inhibitory effects of 0.01% SDS, and 0.1% each Tween 20/NP40 completely reverses the inhibitory effects of 0.01% SDS in the presence of DNA and Mg^{++} (no dNTP) after 40 min at 37°C (S. Stoffel, in preparation).

Taq DNA polymerase shows considerable amino acid sequence similarity to *E. coli* PolI.[7] Significant similarity is observed in regions of the amino-terminal 1/3 of the two enzymes. This domain is known to contain the 5'-3'-exonuclease domain of PolI. Of particular interest is the observation that all of the sites known to be critical for the 5'-3'-exonuclease activity of *E. coli* PolI are perfectly conserved in *Taq* DNA polymerase. Significant identity is also observed for regions of PolI that are involved in 3'-OH primer interaction, dNTP binding, and DNA template binding. In contrast, no meaningful alignment was obtained for the 3'-5'-exonuclease domain of *E. coli* PolI, possibly accounting for the observed lack of 3'-5'-exonuclease activity in *Taq* DNA polymerase.

In addition to the isolation and purification of DNA polymerase from *T. aquaticus*, DNA polymerase activities from *Bacillus stearothermophilus*,[14,15] several other *Thermus* species[16-18], and several divergent archaebacterial species[19-21] have been reported and partially characterized. These organisms were grown at temperatures ranging from 60°C (*B. stearothermophilus*)[15] to 87°C (*S. solfataricus*).[19] DNA polymerases are very susceptible to proteolytic attack and care must be exercised during harvest of the cell material and during cell lysis and purification to minimize the chance of attributing different purification characteristics or enzymatic properties to partial enzyme fragments. The most extensively purified archaebacterial DNA polymerase had a reported 15-min half-life at 87°C.[21] There are several extreme thermophilic eubacteria and archaebacteria that are capable of growth at very high temperatures.[22,23] Some of these organisms may contain very thermostable DNA polymerases, may show a high degree of amino acid sequence similarity to the *E. coli* DNA polymerase I family[7] or to the eucaryotic DNA polymerase family,[24,25] or may represent a distinct class of new DNA polymerases.

FUTURE ISSUES

Aspects of DNA synthesis at high temperature, enzymatic, biochemical and structural properties of *Taq* and other thermostable DNA polymerases, as well as the identification and characterization of replication accessory proteins are all challenging areas for further investigation. To what degree is *Taq* DNA polymerase capable of efficient displacement synthesis at a replication fork? If only limited synthesis occurs under such conditions, are there *Thermus* "helicases" which facilitate efficient displacement synthesis? The ability to catalyze processive, displacement synthesis could ameliorate one of the factors which may contribute to "plateau effect" and limit the final amount of specific product accumulation in a PCR (see Chapter 1). What are the structural features which contribute to the thermostability of *Taq* DNA polymerase? Since the regions and specific amino acid residues of *E. coli* PolI that are known to be critically important for 5'-3'-exonuclease activity, dNTP binding, primer and template interaction are remarkably conserved in *Taq* DNA polymerase, other portions of the enzyme must critically determine thermostability and activity at high temperature. A comprehensive understanding of the enzymatic, biochemical and structural properties of thermostable DNA polymerases is likely to lead to an improved ability to generate very large products as well as improved specificity, final yield of desired product and enhanced sensitivity of detection of rare targets in a PCR.

ACKNOWLEDGMENTS

I thank Chu-An Chang, Susanne Stoffel and Robert Watson for permission to cite their data prior to publication; Corey Levenson, Lauri Goda and

Dragan Spasic for the synthesis of many oligonucleotides; Lynn Mendelman for providing a preprint prior to publication; Will Bloch and members of the Cetus PCR group for stimulating discussions, thoughtful advice and suggestions. I also thank Tom White and Jeff Price for their interest, patience, wisdom and support of these efforts.

REFERENCES

1. Saiki, R.K., Gelfand, D.H., Stoffel, S., Scharf, S.J., Higuchi, R., Horn, G.T., and Erlich, H.A. (1988) *Science* 239:487-491.
2. Mullis, K.B., and Faloona, F. (1987) *Methods in Enzymology* 155:335-350.
3. Saiki, R.K., Scharf, S., Faloona, F., Mullis, K.B., Horn, G.T., Erlich, H.A., and Arnheim, N. (1985) *Science* 230:1350-1354.
4. Brock, T.D., and Freeze, H. (1969) *J. Bacteriol.* 98:289-297.
5. Chien, A., Edgar, D.B., and Trela, J.M. (1976) *J. Bacteriol.* 127:1550-1557.
6. Kaledin, A.S., Slyusarenko, A.G., and Gorodetskii, S.I. (1980) *Biokhimiya* 45:644-651.
7. Lawyer, F.C., Stoffel, S., Saiki, R.K., Myambo, K., Drummond, R., and Gelfand, D.H. (1989) *J. Biol. Chem.* 264:6427-6437.
8. Innis, M.A., Myambo, K.B., Gelfand, D.H., and Brow, M.A.D. (1988) *Proc. Natl. Acad. Sci. USA* 85:9436-9440.
9. Sabatino, R.D., Myers, T.W., Bambara, R.A., Kwon-Shin, O., Marraccino, R.L., and Frickey, P.H. (1988) *Biochemistry* 27:2998-3004.
10. Tindall, K.R., and Kunkel, T.A. (1988) *Biochemistry* 27:6008-6013.
11. Mendelman, L.V., Boosalis, M.S., Petruska, J., and Goodman, M.F. (1989) *J. Biol. Chem.*, in press.
12. Petruska, J., Goodman, M.F., Boosalis, M.S., Sowers, L.C., Cheong, C., and Tinoco, Jr., I. (1988) *Proc. Natl Acad. Sci. USA* 85:6252-6256.
13. Scharf, S.J., Horn, G.T., and Erlich, H.A. (1986) *Science* 233:1076-1078.
14. Stenesh, J., and Roe, B.A. (1972) *BBA* 272:156-166.
15. Kaboev, O.K., Luchkina, L.A., Akhmedov, A.T., and Bekker, M.L. (1981) *J. Bacteriol.* 145:21-26.
16. Kaledin, A.S., Slyusarenko, A.G., and Gorodetskii, S.I. (1981) *Biokkimiya* 46:1576-1584.
17. Kaledin, A.S., Slyusarenko, A.G., and Gorodetskii, S.I. (1982) *Biokkimiya* 47:1785-1791.
18. Rüttimann, O., Cotoras, M., Zaldivar, J., and Vicuna, R. (1985) *Eur. J. Biochem.* 149:41-46.
19. Rossi, M., Rella, R., Pensa, M., Bartolucci, S., DeRosa, M., Gambacorta, A., Raia, C.A., and Orabona N.D-A. (1986) *System. Appl. Microbiol.* 7:337-341.
20. Klimczak, L.J., Grummt, F., and Burger, K.J. (1986) *Biochemistry* 25:4850-4855.
21. Elie, C., DeRecondo, A.M., and Forterre, P. (1989) *Eur. J. Biochem.* 178:619-626.
22. Bergquist, P.L., Love, D.R., Croft, J.E., Streiff, M.B., Daniel, R.M., and Morgan, W.H. (1987) *Biotech. & Genet. Eng. Rev.* 5:199-244.
23. Kelly, R.M., and Deming, J.W. (1988) *Biotechnol. Prog.* 4:47-62.
24. Bernad, A., Zaballos, A., Salas, M., and Blanco, L. (1987) *EMBO* 6:4219-4225.
25. Wang, T.S.-F., Wong, S., and Korn, D. (1989) *FASEB J.* 3:14-21.

CHAPTER 3

PCR Automation

Christian Oste

INTRODUCTION

In the evolution of the Polymerase Chain Reaction (PCR), two developments have greatly simplified the procedure: automation of temperature cycling and the use of a thermostable DNA polymerase. The original method, using the Klenow fragment of *E. coli* DNA polymerase I, was very tedious because of the thermal lability of the enzyme. The initial PCR process required moving the samples to be amplified between two heat sources: one at high temperature (94°-95°C), required to denature the double-stranded DNA, and one at relatively low temperature (37°C), needed for both the annealing of the primers and their extension. Because the Klenow fragment is irreversibly denatured every time the sample temperature is raised to 94-95°C, the enzyme has to be replenished every cycle, after the sample temperature is restored to 37°C, in order to extend the annealed primers. Thus, the requirements of automating the PCR method at the time of its invention were actually two-fold: 1) cycling the temperature between 94°C and 37°C, and 2) adding fresh enzyme every cycle.

Automation was needed, especially for the temperature cycling, but no suitable devices existed at the time PCR was invented to meet this need. Cetus Instrument Systems modified a Pro/Pette™ liquid handler (the Perkin-Elmer Cetus multi-channel automated liquid handler) to function as the first automated system to perform this PCR method. The bed of this instrument, dubbed "Mr. Cycle," was modified to accommodate two temperature controlled aluminum blocks, each with a 48-sample capacity. The front block held the samples (in uncapped microcentrifuge tubes) and was connected, via a switching valve, to two water baths, one at 94°C, the other at 37°C. The back block held solutions of Klenow fragment, also in uncapped microcentrifuge tubes and placed in the same configuration as the samples in the front block. To ensure the stability of the enzyme solutions, the back block was connected to a water bath set at 4°C. A controller kept track of the incubation times at both high and low temperatures, actuated the switching valve in order to change the front block temperature and prompted the multi-channel head to pick up fresh tips, withdraw aliquots of enzyme solution from the tubes in the back block and deliver them into the corresponding samples in the front block.

Although this instrument worked reasonably well, it was clear that a new approach could simplify the automation of PCR. This second advance came from purification of a highly thermostable DNA polymerase to perform the PCR. As reported elsewhere in this book, the DNA polymerase from *Thermus aquaticus* (*Taq* DNA Polymerase) needs to be added but once to the sample at the beginning of the procedure; usually, no further addition is required, thereby completely eliminating the need for liquid handling.

Several other benefits flowed from this change in the enzyme. Because of its relatively high (70-75°C) temperature optimum, the typical temperature profile for a cycle of PCR using *Taq* DNA polymerase could now feature three distinct temperatures instead of the two used with the Klenow fragment: 1) a denaturation step, with temperatures in the range of 92-97°C, 2) a primer annealing step, at temperatures above 37°C, and 3) a primer extension step, usually at temperatures between 65°C and 80°C, with an optimum centered around 72°C. Now the temperatures for the annealing and extension steps could be optimized, independently. One consequence was a gain in specificity due to the higher annealing temperature. Although the need for complex liquid handling was eliminated, the requirement for accurate control of the temperature became more important.

AUTOMATED INSTRUMENTS

Two types of approaches[1-3] to automating the PCR process with the *Taq* polymerase can be envisioned, each presenting some advantages and disadvantages.

1. Robotics: the temperature control stations can be stationary and the samples are moved between them in a mechanized duplication of the

manual operation, according to the requirements for each incubation step.

2. Temperature cycling devices: the samples remain in the same physical position in the sample holder throughout the process, while a software-controlled heating/cooling system is responsible for effecting the regulated, cyclic temperature changes and incubations.

The first approach, typically robotic, has been used in a variety of formats, from rudimentary to relatively sophisticated. In most variations, the samples undergo a two-dimensional translation from one station to the next. However, at least one system operates on the principle of a lazy-Susan, where the samples are lifted vertically and the next station is moved underneath the sample holder, prior to lowering the samples. Systems using an arc need the stations to be spread further apart, requiring more bench space. A linear pattern allows the temperature control stations to be placed next to each other, reducing the bench space needed. The robotic arm is often mounted on a rail, although the stations can also be placed under the span of a "loading crane" type of device, as used in some automated liquid handlers.

A circular motion is likely to lead to progressively accumulating, positioning errors. So, although a unidirectional motion, describing a circle, may be easier to control and to reproduce accurately, it cannot be recommended in combination with a system of stations requiring accurate positioning of the sample, as would be required with "dry" heating/cooling blocks. Practically, the best option is to use baths filled either with water or with oil, since they offer more forgiving landing zones for the sample holder. The least sophisticated system of this type which I am aware of consists of three polystyrene buckets, filled with water whose temperature is regularly readjusted manually, and an arm moving along a circular path but devoid of vertical displacement. The sample holder is a cup, perforated at the bottom in order to fit the samples tubes. Since the arm does not have a 2-axis motion, the holder has to tilt as it passes over the rim of each bucket-stations. In general, this type of robotic system is usually simple to set up; the different components are available right off the shelf. Also, very little software is needed, since the options for control are so limited by the physical system.

In summary, the disadvantages of a robotic system to perform the PCR process are several. First the footprint of such a system is usually quite large, regardless of the type of motion effected by the arm. A normal laboratory bench may not be deep enough to accommodate a whole set-up, except in the case of some linear displacement systems. Second, fine-tuning such a system may be a lengthy process, as water baths and heating/cooling blocks are notorious for slow temperature equilibration, even after a small temperature change is dialed in. Third, control of the rate of temperature change in the samples is impossible during the transport through air. This becomes particularly important in the case of the transition between annealing and extension steps. When the temperature of the sample may in fact dip below the value initially optimized for the annealing step, possibly reducing specificity.   Finally, but perhaps most significantly, the options for control of temperature and segment lengths are highly constrained.   Some investigators find changing temperatures from cycle to cycle can improve results. For these reasons, the robotic approach, although attractive because of its intrinsic simplicity, is less likely to win over the typical PCR user than the alternative: stationary samples and self-contained temperature regulation system.

The second approach is a fully integrated, automated and dedicated PCR system.   A variety of architectures have been developed either as new devices for PCR amplification or as significant modifications of already existing laboratory instrumentation.   The ancestor of all such PCR-dedicated instruments was developed by Cetus Engineering in 1985-86, after it was established that *Taq* DNA polymerase could be efficiently used in the PCR procedure.   This instrument was nicknamed "Son-of-Cycle" and used semi-conductor elements (Peltier devices) both for heating and cooling the sample holding block.   An off-the-shelf controller regulated the cycling and monitored both incubation times and block temperature.   This system works quite well (and is still being used occasionally at Cetus), but is limited by the progressive wearing of the Peltier devices.   The thermal stress across the junctions in the Peltier devices during many cycles eventually leads to the rupture of the junctions, disabling them completely.   Having to replace the Peltier devices every 3-6 months, depending on frequency of usage, would not be acceptable to most users.   A second problem is the poor energy efficiencies of Peltier devices, which slows the cooling to marginally acceptable rates.

A number of architectures have been explored in the design of dedicated PCR systems.   They basically fall into three categories, the samples remaining stationary in all cases:

1.   heating and cooling by fluids

2.   heating by electric resistances/cooling by fluid (open or sealed loop, liquid or gas)

3.   heating by electric resistances/cooling by semi-conductors

The second concept, using a combination of heating by electrical resistances and cooling by fluid, was also the second design chosen and has found its way into a variety of commercially available instruments. Only one of them is truly self-contained, using a closed refrigeration circuit for cooling, as opposed to relying on tap water. This concept lends itself to higher levels of sophistication in temperature control. For instance, using electrical resistances for heating the sample holding block should, in theory, allow for controlled rate of temperature change, which appears to be critical in the case of very short primers. Yet, at this point, only one offering actually is capable of accurately modulating the ramping rate, both in heating and cooling modes. (It should be pointed out that the modulation in the cooling mode is possible only because that instrument has a built-in refrigeration system and doesn't rely on outside source such as tap water for cooling the block).

Instruments using tap water for cooling the block present some disadvantages. First, most of them require a fairly high minimum water pressure in order to overcome the back pressure from the filtering device fitted on the inlet port. If, for some reason, this minimum pressure at the tap is not reached, some pressure will however start building up in the tubing that could result into some extensive water-induced damages in the general laboratory environment. Also some residual tap water will always stagnate inside the block at the end of the experiment, which may result in a progressive (and invisible) corrosion of the block. Second, the temperature of tap water is normally subject to seasonal variations, not to mention geographical variations, meaning that, even within the same laboratory, cooling rates may vary on a seasonal basis, possibly affecting the efficiency of the primer-template complex formation and therefore the overall amplification efficiency. Third, washing tap water directly down the drain, just to cool the sample holding block, may, in many parts of the world, be considered unacceptable from an ecological standpoint. The alternative would be to connect the instrument to a refrigerating water bath, *de facto* recycling the cooling fluid. This, by the way, is also the only way sub-ambient temperatures can be achieved in the sample. Considering the high cost of refrigerating water baths, and the increase in bench space required to install it, it appears that the instrument with built-in, closed loop refrigeration circuit is a more rational alternative. The capability for the instrument to reach sub-ambient temperatures represents an important added advantage towards fully automated, unattended operation, provided that the software is simultaneously capable of executing several programs in succession. Indeed, under those circumstances, it will be possible to lower the samples temperature to 4-6°C (refrigerator temperatures), thereby improving their safekeeping until the operator actually retrieves them. (These devices can also be used for other purposes such as restriction digests, blunt-end ligations, nick-translations, etc.) In summary, although conceptually relatively simple, dedicated instruments using an external source of cooling, such as tap water, may present a number of disadvantages which may affect the ease of use of the instrument as well as the practical reproducibility of the experimental results.

In contrast, by using a combination of electrical resistances for the heating and semi-conductors for the cooling, it becomes possible to build a more fully integrated system than by relying on external source of cooling. In particular, if the Peltier devices come in contact with the sample holding block only during the cooling step (from denaturation to annealing), the stress on the junctions will be reduced although never completely eliminated. At the beginning of the cooling step, one of the sides of the elements will be instantaneously brought from low sub-ambient temperature to 94-95°C, while the other side remains at the low temperature. This rapid temperature transition generates a significant amount of stress on the solder at each junction between the two sides of the Peltier devices. Eventually, as mentioned earlier, the solder will progressively become more brittle, slowly impairing the whole device. The commercial unavailability of semi-conductors elements which use stress-resistant solder was in fact the main reason why this concept was abandoned in the first prototypes developed at Cetus. Because the failure of the Peltier devices is never catastrophic, the reduction in instrument performance would not become readily noticeable, possibly affecting the experimental results unbeknownst to the user. At any rate, this concept may be more acceptable and efficient than an external cooling system, provided that its limitations and potential reliability problem are understood.

PARAMETER CONTROLS

Three additional points, valid for all dedicated PCR systems, need to be addressed: 1) the software needed for fully automated operation, 2) the accuracy and reproducibility of temperature control in all samples processed simultaneously, and 3) the formatting of the reaction.

In order to fully automate the operation of a dedicated PCR system, it is obvious that the software controlling the instrument will have to be capable of 1) accurately and reproducibly controlling incubations and transitions, including the capability of slowing the latter down if needed, 2) execute several independent routines in sequence, 3) store in a non-volatile memory whole procedures (see #2), 4) output, in real-time, valuable information about the actual performance of the instrument, both in incubations and transitions. In addition, direct access to the parameters to be programmed via intelligible prompts is of enormous advantage to the user, because it reduces both time and confusion at the level of program editing.

To guarantee the success of the experiment and allow the user to compare directly samples from the same run, or, on the long-term, samples run at different times using the same program, it is of paramount importance that all samples reach the same temperature during each incubation. This is a problem specific to sample holding blocks rather than to all-fluid type of configurations. Two main criteria should be considered in the case of blocks: 1) the homogeneity of cooling and heating across the whole block, and 2) the physical fit between the wells and the microcentrifuge tubes containing the samples.

More than anything else, those two criteria address specific mechanical engineering problems, such as proper choice of alloy for the block, precision milling of the wells, minimization of the total thermal mass, etc., which go beyond the scope of this chapter. As a method of reporting the actual sample temperature to the software, the use of a thermocouple in a mock sample in one of the wells has two major drawbacks: 1) most thermocouples used for that purpose have a significant thermal mass *per se* and will, therefore, consistently report lower temperatures than actual, and another drawback is that thermocouples degrade over time, especially when left in water, 2) if the block temperature is not homogenous to start with, depending on where the probe is located, the whole PCR profile may end up being biased towards higher or lower temperatures.

A better approach consists of having sensors monitor the block temperature continuously, feed the raw measurement data into an algorithm that will mimic the actual temperature at the level of the sample (in the tube) and instruct the control software to take corrective action by heating or cooling according to needs. This is a sophisticated approach to temperature control and it will work best with a block that is thermically homogenous from the start. This is also the only method that will actually take into account the lag period normally encountered during transitions, i.e., when the sample temperature change is somewhat delayed *vis-a-vis* the block temperature change. The result of using this method of temperature control is that temperature overshoots, which could prevent amplification, and undershoots, which could reduce the specificity of the reaction, are virtually eliminated. In summary, accurate temperature control and temperature homogeneity across the block are two very important criteria in the success of a PCR experiment.

The question of the formatting of the samples leads to considering satellite instrumentation. Indeed, the current practice of setting up the samples in microcentrifuge tubes requires multiple pipetting steps, which could aggravate the carryover problem that has already been observed in PCR. Although it is possible to partially automate the mixing of the different components of the PCR reaction, it is obvious that switching over from a tube format to a microtiter plate-like format may present some advantages. It is conceivable that, at that point, the sample preparation could actually be performed by an automated liquid handler. However, the choice of proper material for the reaction vessel will be of critical importance: 1) the vessel should not warp when exposed to high temperatures, thereby ensuring a constant and homogeneous thermal transfer during incubations and transitions, 2) it should not leach organic compounds that could inhibit the polymerization reaction, 3) the temperature should be homogeneous across the vessel at all times, and 4) monitoring the vessel temperature should be as easy and accurate as with a sample holding block. Another design consideration is to minimize the potential for sample-to-sample cross-contamination in the shallow wells.

Although there is currently one instrument that offers the capability of running the PCR reaction in a microtiter plate, its performance remains in part questionable, mostly because currently available microtiter plates do not conform to the criteria listed above. It is clear that a format allowing for more automation at the pre-PCR level will be favored by most users in the future because that same type of format may, in fact, also facilitate the post-PCR analysis, be it by direct detection of the PCR product or by allowing for the easy withdrawal of sample aliquots which, in turn, will be engaged in subsequent reactions such as sequencing.

SUMMARY

In conclusion, the PCR methodology offers an exciting opportunity for new instrumentation development. At this point, the sample preparation and analysis have actually become the bottlenecks in the procedure. Indeed, given the proper dedicated PCR system, performing the PCR reaction itself has now become a routine procedure. Some criteria remain very important for the success of the experiment, such as accuracy and reproducibility in temperature control, software flexibility, small footprint and full integration of the system. It is likely that efforts will be put into improving performance in transition and flexibility in sample formatting. In addition, one should also expect to see significant advances in areas dealing with sample preparation (pre-PCR) and analysis (post-PCR). Some instrumentation manufacturers are already addressing those areas with the hope of developing a fully compatible, possibly modular bench-top, system that would handle the samples from start to finish with minimal operator interaction.

REFERENCES

1. Foulkes, N.S., Pandolfi de Rinaldis, P.P., McDonnell, J., Cross, N.C.P., and Luzzatto, L. (1988) *Nucl. Acids Res.* 16:5687-5688.
2. Kim, H.S., and Smithies, O. (1988) *Nucl. Acids Res.* 16:8887-8903.
3. Oste, C. (1988) *BioTechniques* 6(2):162-167.

CHAPTER 4

Simple and Rapid Preparation of Samples for PCR

Russell Higuchi

INTRODUCTION

Sample preparation for PCR can be as simple and rapid as adding cells directly to the PCR.[1] Things are not quite this easy, however, for all samples and applications of PCR. In this chapter, I will discuss what determines the amount of preparation necessary and present example protocols for the rapid preparation for PCR of DNA and RNA from various sources.

In the simplest case cited above, enough of the DNA from a small number of tissue-culture cells was made accessible to PCR merely by the lysis of the cells during the heat-denaturation step. Because of the ability of PCR to produce detectable amounts of product even from a few genome equivalents of DNA,[2] it is not absolutely necessary that making the DNA available for PCR be a particularly efficient process. However, there are several reasons why one would prefer it to be. These are: 1) the more template molecules available, the less likely are false results due to either cross-contamination between samples or "carryover" of analogous PCR product from earlier amplifications, 2) if the PCR amplification is not particularly specific or efficient, or the target sequences are present infrequently among a large number of cells (see below), the yield of product will be inadequate without enough starting DNA, and 3) it is more difficult to infer target DNA content from PCR product yield when the fraction of the starting DNA available to PCR is uncertain.

SAMPLE PREPARATION

For the direct, cell-lysis protocol, the easiest way to increase the number of templates available to PCR should be to increase the number of cells added to the reaction.   As shown in Figure 4 of Saiki *et al.*,[2] however, as the number of cells added goes above 600, the yield of product, unlike the yield from an equivalent amount of purified genomic DNA, does not increase with the addition of more cells, and begins to decline at 4800 cells. Due to the observed "cell-debris," we infer that either more and more DNA is being trapped or that inhibition of the PCR process is occurring or both.

For some applications, this limit on available template is unacceptable. This is true for the screening of samples for infectious agents such as HIV, which occurs in only a few white blood cells out of many thousands.   One needs to efficiently screen all the DNA from tens of thousands of cells in order to pick up the HIV DNA from the few infected cells.   Other applications where this is important include the screening of transgenic organisms for the transgene,[3] if the transgene is present in only a minority of the cells of a tissue, and the monitoring of bone-marrow transplants for the relative amounts of donor and recipient cells over time.

The problem, then, is to find conditions that simultaneously release DNA and/or RNA from larger numbers of cells in a form suitable for PCR while preserving the activity of *Taq* DNA polymerase.   Methods of DNA purification from animal cells often use detergents to solubilize cell components and a proteolytic enzyme to digest away proteins, probably mainly histones, that would otherwise remain strongly bound to the DNA. This procedure is usually followed by extraction with organic solvents to remove residual proteins and membrane components followed by steps, such as precipitation of nucleic acids by ethanol, to remove traces of the organic solvents.   Since *Taq* DNA polymerase activity is not significantly affected by certain non-ionic detergents,[4] and since Proteinase K could be inactivated by heat, cells were added directly to a PCR containing non-ionic detergents and Proteinase K, but not yet containing *Taq* DNA polymerase.   The Proteinase K was given time to work, and after the residual proteinase activity had been destroyed by incubation at 95°C for 10 min, *Taq* DNA polymerase was added and amplification cycles begun.

As shown in Figure 1, comparative yields of PCR product were obtained from up to 70,000 cells or the equivalent amount of purified DNA using this procedure (see Protocol A, end of chapter).   This protocol is routinely used to screen 300,000 cells at a time in an HIV detection assay[5] (S. Kwok, personal communication).   Evidently there is little inhibition of PCR by the

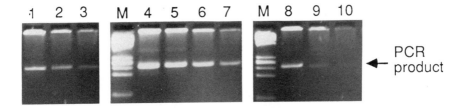

Figure 1. Amplification of class II HLA-DQα sequences[2] from whole blood (lanes 4-7) prepared using Protocol B; density gradient purified mononuclear cells (lanes 8-10) prepared using Protocol A or from purified DNA (lanes 1-3). The number of nucleated cells added were: 250,000, 125,000, 60,000, and 30,000 (lanes 4-7, respectively; 67,000, 33,500, and 16,750 (lanes 8-10, respectively). Purified DNA in amounts equivalent to that from 67,000, 33,500, and 16,750 cells was amplified and run in lanes 1-3. Twenty-five cycles of PCR were performed, 10% of each PCR was electrophoresed on this NuSieve (FMC) agarose gel. M is 0.5 μg of øX 174 DNA cleaved with HaeIII.

remaining cell components and most of the DNA is available to be amplified. This procedure has worked with washed tissue culture cells and density gradient purified (and washed) peripheral mononuclear cells.

If the same procedure is attempted with whole blood rather than purified mononuclear cells, inhibition of PCR occurs with the addition of as little as 1 μl blood to a 0.1-ml reaction. There is a noticeable precipitate in the reaction tubes and purified DNA "spiked" into these samples is not amplified. The testing of various blood components indicates that porphyrin compounds derived from heme may be the most inhibitory substances found in blood. Hematin has been found to inhibit PCR at as low a level as 0.8 micromolar (Walsh and Higuchi, unpublished observation).

To separate quickly porphyrin compounds from nuclear DNA, Protocol B depends on the osmotic lysis of cells and the pelleting of nuclei and cell debris. Hemoglobin released from RBCs is washed away in several pelleting and washing steps. The final resuspension of the pellet containing nuclear DNA is in a Proteinase K/detergent solution similar to that used in Protocol A; from this point the two protocols are essentially the same.

Figure 2 shows the results of an experiment using Protocol B in which a few fibroblasts containing transfected neomycin resistance genes are added to whole blood and detected at dilutions up to 1 per 10,000 nucleated blood cells; a total of 100,000 nucleated cells in whole blood (about 13 μl) were initially added. 300,000 cells have been successfully screened in this manner.

A simpler protocol has been published that uses boiling to simultaneously lyse cells, release DNA, and precipitate hemoglobin.[6] DNA found in the

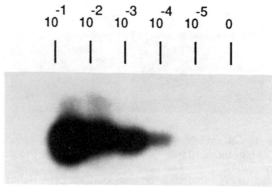

Figure 2. Fibroblast cells transfected with a bacterial neomycin resistance gene (C. Perez, personal communication) were added to whole human blood at a ratio of 1/10, 1/100, 1/1000, 1/10,000, and 1/100,000 nucleated cells. The preparation of the blood was as per Protocol B. The amount of the final lysate sampled was equivalent to the DNA from 100,000 cells. PCR was performed to detect the neomycin resistance gene. Shown is the autoradiograph of a Southern blot performed upon the PCR product and probed with a ^{32}P-labeled neomycin resistance gene probe.

supernatant is added directly to a PCR. However, both the amount of DNA released to the supernatant and the amount of supernatant that can be added to a PCR without inhibition is limited. As discussed above, an upper limit on the amount of template DNA may not be suitable for some applications.

For the preparation of DNA for PCR from tissues other than blood, these same principles apply. Given the amplification potential of PCR, a small amount of DNA can be adequate. Thus, crude lysates can be used by adding small enough amounts that inhibition is avoided. However, if screening for targets that are not present in every cell is required or if the PCR system used is not efficient, then additional effort may be necessary to make available more template DNA. Protocol C below is such a protocol, using non-ionic detergents and Proteinase K as above, that can be used to screen for viral infections in epithelial cells collected as clinical swabs. Protocol D is a protocol for use on single, plucked hairs - a convenient easily obtained source of DNA.

The rapid preparation of mRNA for amplification by PCR has different requirements. The preparation must also be compatible with reverse transcriptase, which is used to make the amplifiable cDNA template, and RNA degradation by endogenous RNAse should be obviated. Protocol E uses DEP to inactivate RNAse and the non-ionic detergent NP40 to lyse cells without disrupting nuclei. mRNA is separated from DNA by centrifugation of the nuclei. DEP can be dissipated by heat and both reverse transcriptase and *Taq* DNA polymerase activities are compatible with NP40. To further ensure that mRNA, and not genomic DNA sequences, are amplified, primers for PCR are usually chosen that are specific to exons separated by at least one intron, such that the product amplified from a genomic DNA template would be of a much larger size, preferably so large as to not amplify efficiently.

PROTOCOLS

Solutions

PBS
- 0.85% w/v NaCl, 66 mM $NaPO_4$ (pH 7.0)

PCR buffer w/nonionic detergents and Proteinase K
- 50 mM KCl
- 10 mM Tris-HCl (pH 8.3) 2.5 mM $MgCl_2$
- 0.1 mg/ml gelatin
- 0.45% NP40
- 0.45% Tween 20

Autoclave and store frozen. When ready to use, thaw and add 0.6 µl of 10 mg/ml Proteinase K (in H_2O) per 100 µl of solution.

Lysis buffer*
- 0.32 M sucrose
- 10 mM tris-HCl (pH 7.5), 5 mM $MgCl_2$
- 1% Triton X-100

Isotonic high pH buffer
- 140 mM NaCl, 10 mM tris-HCl (pH 8.0), 15 mM $MgCl_2$

  *from Buffone, G.J. and Darlington, GJ. (1985) *Clin. Chem.* 30(1):164-5.

Protocol A. Mononuclear Cells Purified from Blood on Ficoll-Hypaque Gradients, or Tissue Culture Cells[7]

If RBCs are <10% of cells, use this procedure. If there is >10% RBC contamination, see below.

1. If <5 ml of cells, bring cells to 10 ml with phosphate buffered saline (PBS) in Falcon #2099, 15-ml conical centrifuge tube, or comparable tube.

2. Centrifuge 100 x G for 2-5 min. Remove supernatant with pipet.

3. Resuspend cells in 10 ml PBS and repeat centrifugation/wash

4. Resuspend in a volume of "PCR buffer w/ nonionic detergents and Proteinase K" to give about $6x10^6$ cells per ml. Transfer to 1.5-ml Eppendorf microcentrifuge tube.

5. Incubate at 50-60°C for 1 hr.

6. Incubate at 95°C for 10 min to inactivate the proteinase.

7. Store frozen.

To effect PCR on the lysate produced in step 6, Protocol A, note that 25 μl of this lysate is about equivalent to one microgram of genomic DNA. Add this lysate in up to half the reaction volume of a PCR to a solution that is one-half the volume of the PCR and 1x in PCR buffer, and 2x in triphosphate, primers, and enzyme. Make up any deficiency in volume with water.

With RBC contamination, do one PBS wash as above, then resuspend in 1 ml of "lysis buffer," as in Protocol B below, and transfer to a 1.5-ml Eppendorf centrifuge tube. Do one spin, as below, remove supernatant and resuspend in "PCR buffer w/noninonic detergents and Proteinase K" (to give about $6x10^6$ cells/ml) and proceed as in step 4 above.

Protocol B. Whole Blood[7]

This procedure dissolves the cytoplasmic membrane and pellets nuclei. Therefore, cytoplasmic DNA is lost. Should yield about 20 μg of DNA.

1. Mix 0.5 ml blood with 0.5 ml "lysis buffer" in a 1.5-ml Eppendorf microcentrifuge tube.

2. Centrifuge 13,000 x G for 20 sec.

3. Remove supernatant with pipet and resuspend pellet, using vortex mixer, in 1.0 ml of "lysis buffer."

4. Repeat steps 2 and 3 twice.

5. Centrifuge 13,000 x G for 20 sec, remove supernatant, and resuspend in 0.5 ml of "PCR buffer w/nonionic detergents and Proteinase K."

6. Follow steps 5-7 above.

Protocol C. Clinical Swabs (M. Manos, Cetus Corp.)

1. Collect cervical, vulvar, or penile samples with a pre-wet (PBS) swab or cytobrush.

2. Place the swab into 2 ml PBS (with 2x concentration of Fungibac; Gibco) in 10- to 15-ml conical tube (Falcon #2099) or comparable. Sample can be kept at room temperature for 24 hr; if longer storage required, keep at 4°C.

3. Remove the swab and centrifuge tube for 5' at 2-3,000 rpm in clinical centrifuge to pellet cells. Remove supernatant by aspiration.

4. If red blood cells are present, resuspend in 1 ml lysis buffer, transfer to a 1.5-ml microcentrifuge tube and proceed as in step 2 of Protocol B above. The DNA concentration in the final resuspension depends on the number of cells present.

5. If no red blood cells are present, resuspend pellet in PCR buffer with nonionic detergents and Proteinase K as in Protocol A, step 4 (use 1% Laureth 1Z instead of NP40 and Tween 20 and Proteinase K at 200 µg/ml). Proceed from there through step 7.

Protocol D. Plucked Hairs

1. Cut off 0.5 cm of freshly plucked hair at root end. Use fine-tipped forceps and razor blade. A dissection scope may be helpful.

2. Place this 0.5-cm piece into 0.4 ml of PCR buffer with nonionic detergents and Proteinase K in 1.5-ml microcentrifuge tube.

3. Proceed as in Protocol A, steps 5-7. Use 50 µl lysate to effect PCR.

Protocol E. RNA from Blood Cells (E. Kawasaki, Cetus Corp.)

1. Prepare mononuclear cells from 1-2 ml blood by Ficoll-Hypaque or similar method.

2. Place mononuclear cells in 2-ml screwcap microcentrifuge tube; fill with PBS and pellet cells at 500 xG for 5 min.

3. Meanwhile, prepare DEP (diethylpyrocarbonate; Sigma - warning: causes irritation, take proper precautions) solution by diluting 1:9 in absolute EtOH then 1:999 in IHB that is 0.5% in NP40. Keep ice-cold.

4. Resuspend cell pellet in 200-400 µl of this solution. Vortex to mix.

5. Pellet nuclei by centrifugation at 13,000 xG for 10 sec.

6. Transfer supernatant to new tube and incubate at 37°C for 20 min, then 10 min at 90°C; keep the cap loose to allow gases produced from DEP to escape.

7. Pellet any precipitate that forms and transfer supernatant to new tube. Use 5-10 µl in reverse transcriptase (RT) reaction as described in Chapter 7.

8. If RT reaction fails, it may be because of residual DEP. Reheat sample at 90°C for 5 min and try again.

9. This method can be used for tissue culture cells as well, but use 5-10 x fewer cells as tissue cells have that much more RNA per cytoplasm.

REFERENCES

1. Saiki, R.K, Bugawan, T.L., Horn, G.T., Mullis, K.B., and Erlich, H.A. (1986) *Nature* 324:163.
2. Saiki, R.K., Gelfand, D.H., Stoffel, S., Scharf, S.J., Higuchi, R., Horn, G.T., Mullis, K., and Erlich, H.A. (1988) *Science* 239:487.

3. Tsukumoto, A.S., Grosschedl, R., Guzman, R.C., Parslow, T., and Varmus, H.E. (1988) *Cell* 55:619.
4. Lawyer, F.C., Stoffel, S., Saiki, R.K., Myambo, K., Drummond, R., and Gelfand, D.H.   *J. Biol. Chem,* in press.
5. Ou, C., Kwok, S., Mitchell, S.W., Mack, D.H., Sninsky, J.J., Krebs, J.W., Feorino, P., Warfield, D., and Schochetman, G. (1988) *Science* 239:295.
6. Kogan, S.C., Doherty, M., and Gitschier, J. (1987) *N. Engl. J. Med.* 317:985.
7. Higuchi, R. (1989) Perkin Elmer/Cetus Newsletter *Amplifications* 2:1.

PART TWO

RESEARCH APPLICATIONS

The ability to synthesize large amounts of a specific DNA fragment from a complex template has significantly facilitated subsequent analysis. As discussed in Chapter 5, the nucleotide sequence of amplified DNA fragments can be determined directly without molecular cloning and preparation of template by growth of the host and biochemical purification of the vector. The study of sequence-specific interactions of proteins and the DNA fragments generated by PCR is discussed in Chapter 6 as is the ability to introduce specific mutations for functional analysis.

Although nucleotide sequence determination is the most direct and comprehensive method of analyzing genetic variation in the amplified DNA, a variety of indirect methods for detecting mutations and/or sequence polymorphisms have been developed and applied. One of the most general and powerful approaches, the use of denaturing gradient gel electrophoresis (DGGE)[1] to identify sequence differences between amplification products is described in Chapter 7. Differences in the melting profiles of different sequences are reflected in the differential mobility of the amplified fragments. To enhance the resolution of DGGE, a high-melting domain

("GC clamp") is incorporated into the amplified fragment via the 5' end of one of the PCR primers. An alternative method that has been reported recently is RNase A cleavage,[2] an approach in which a labeled RNA probe is hybridized to the amplified fragment and a mismatch revealed by identification of the RNAse A cleavage product by gel electrophoresis. A chemical method of cleaving probe-target duplexes at mismatched nucleotides has also been described.[3] The incorporation of a nucleotide analog which causes a shift in electrophoretic mobility has also been used to detect sequence differences in the amplified products.[4]

These methods are all reasonably general strategies for detecting unknown mutations or sequence polymorphisms; the location and nature of the substitutions must obviously be determined by sequencing. Of course, once an informative sequence variant has been identified, it can be detected using sequence-specific oligonucleotide hybridization probes or amplification primers (see Part Three).

Although not a general approach, restriction site analysis of amplified DNA[5,6] remains a valuable method for detecting genetic variation in some cases. Length polymorphisms in the amplified product that result from variable number tandem repeats (VNTR) in the template have also been used as informative genetic markers.[7,8] Here, the primers hybridize to unique sequences flanking the repeated element. In some cases, like the $(dC-dA)_n$. $(dG-dT)_n$ repeat, the VNTR region is found throughout the genome. The ability to amplify these short repeats and resolve the length variants by gel electrophoresis[9] makes available for genetic analysis a large, previously inaccessible, set of markers. Other VNTR regions that have been analyzed by PCR are single-locus markers with many alleles.[8]

In addition to its use in analyzing genetic variation, PCR is a powerful method for detecting gene expression by synthesizing a cDNA PCR template from an mRNA transcript with reverse transcriptase. This approach, discussed in Chapter 8, has been applied in a variety of studies to the detection of rare mRNA molecules in specific cell lineages.[10] Recently, methods for quantitating the level of specific mRNA transcripts have been reported.[11] The ability to detect in easily accessible tissue or cells the basal level of specific mRNA transcripts for genes normally expressed in other tissues was recently demonstrated.[12] This approach, which should facilitate the sequence analysis of mRNA transcripts expressed in tissues that are difficult to obtain, involves the synthesis of RNA copies of amplified DNA fragments which contain phage promoters incorporated into the primers.[12] By introducing a translation initiation signal via the appropriate primer, *in vitro* translation of these RNA copies can be carried out.[12]

In PCR, the capacity to specifically amplify a particular DNA segment results from the use of two primers whose sequence is complementary to nucleotide sequences flanking the target region. Thus, only DNA segments with sequence information available for primer design can be *specifically* amplified. While this still allows the analysis of mutations, polymorphisms, and evolutionary changes in the sequences of known genes, this requirement represents a constraint on the use of PCR to analyze uncharacterized DNA.

This constraint can be overcome in a variety of ways. For *unknown* DNA sequences where no sequence information is available to serve as the oligonucleotide primer sites, the solution is to create primer sites by adding flanking DNA of a known sequence. The first example of this approach was the amplification of unknown cDNA sequences cloned into a λgt11 vector by using primers specific for the vector sequences flanking the insertion site.[13] In general, flanking sequences for priming can be added by ligation[14] or by homopolymer tailing with terminal transferase.[15] In some cases, a degree of specificity is maintained by using one primer complementary to a target-specific sequence and another primer that anneals to a site created by ligation or terminal transferase tailing. These approaches, termed "anchored PCR,"[15] "RACE,"[16] and "one-sided PCR,"[17] have been applied to cDNA and bacterial genomes, which require a less specific amplification procedure since the template is less complex than human genomic DNA. The combination of a specific primer for the constant region of the T cell receptor β genes and an "anchored tail" as a non-specific primer site has proved very successful in the analysis of the T cell receptor mRNA transcripts.[15] The study of the expressed repertoire of T cell receptor genes has already yielded significant insights into the specificity of the pathogenic autoimmune response in animal models.[18,19]

The use of a degenerate pool of primers based on protein sequence, discussed in Chapter 9, has been used to amplify cDNA sequences.[20] If the amino acid sequence is an evolutionarily conserved motif in a particular protein (e.g., the reverse transcriptase of retroviruses), such degenerate primer pools can be used as a powerful strategy for identifying unknown members of a given virus family.[21] By adjusting the reaction conditions and the position of the mismatch between the primer and the template, one can create either a "tolerant" PCR (for degenerate primer pools) or a "stringent" PCR (for allele-specific amplification).

In general, for all these strategies for the amplification of unknown sequences, there is a "trade-off" between selectivity and amplification. In some strategies for "generic" or "universal" amplification using the ligation of primer sites, the selectivity is introduced independently of the amplification. One study reported the amplification and cloning of sequences in a micro-dissected chromosomal region.[22] Another recent report describes the ligation of primer sites to genomic DNA fragments ("whole genome PCR"), the selection of those fragments that bind a specific protein, and the subsequent amplification and cloning of the enriched sequences.[14] In "*Alu* PCR," discussed in Chapter 11, primers specific for the human *Alu* repeat allow the selective amplification of human sequences from hybrid cell lines containing both human and rodent genomic DNA.

Another elegant modification of the PCR method known as "inverse PCR" or "inside-out" PCR has been developed for the analysis of sequences that flank a known region.[23,24] In this approach, discussed in Chapter 10, the template is digested with a restriction enzyme that cuts outside the region of known sequence, the resulting fragment circularized by ligation, and amplification carried out with primers whose 3' ends point away from each other. Thus, the initial flanking sequences that become joined by ligation lie *between* the 3' ends of the primers and can, therefore, be amplified and analyzed.

The identification of very rare DNA sequences made possible by PCR has opened up new avenues of research. The detection of rare events like the targeted modification of specific genes by homologous recombination is uniquely suited to the PCR method since a "donor" specific primer and a "recipient" specific primer can be used to uniquely amplify a recombinant sequence.[25] This same approach can be applied to the detection of specific translocations, deletions, and insertions by the amplification of novel "junctional" sequences.

The ability of PCR to amplify informative sequences from a single template has important implications for the analysis of genetic recombination and the construction of genetic maps.[26] The co-amplification of sequences from two (or more) linked loci from individual gametes represents a novel and powerful approach to gene mapping, discussed in detail in Chapter 12, that is independent of the analysis of genetic crosses and pedigrees. The construction of phylogenetic trees based on comparing the sequence of the homologous gene in many different species has also been significantly facilitated by PCR and is discussed in Chapter 13. Thus, in the few years since the development of PCR, it has contributed broadly to the molecular analysis of genetic variation, expression, recombination, and evolution.

REFERENCES

1. Fischer, S.G., and Lerman, L.S. (1983) *Procl Natl. Acad. Sci. USA* 80:1579-1584.
2. Almoguera, C. *et al.* (1988) *Cell* 53:549-554.
3. Cotton, R.G., Rodriguez, N.R., and Campbell, R.D. (1988) *Proc. Natl. Acad. Sci. USA* 85:4397-4401.
4. Trainer, G.L., *et al.* (1989) *J. Cellular Biochem.* S13E, 289.
5. Mullis, K.B., and Faloona, F. (1987) *Meth. Enzymol.* 155:335.
6. Kogan, S.C., Doherty, M., and Gitschier, J. (1987) *New Engl. J. Med.* 317:985-990.
7. Jeffreys, A.J., Wilson, V., Newmann, R., and Keyte, J. (1988) *Nucl. Acids Res.* 16:10953-10971.
8. Horn, G.T., Richards, B., and Klinger, K.W. (1989) *Nucl. Acids Res.* 17:2140.
9. Weber, J.L., and May, P.E. (1989) *Am. J. Hum. Genet.* 44:388-396.
10. Rappolee, A.D., Mark, D., Banda, M.J., and Werb, Z. (1989) *Proc. Natl. Acad. Sci. USA* 241:708-712.
11. Gilliland, G., Perrin, S., and Franklin, H. (1989) *J. of Cellular Biochem.* S13E:270.
12. Sarkar, G., and Sommer, S. (1989) *Science* 244:331-334.
13. Saiki, R.K., Gelfand, D.H., Stoffel, S., Scharf, S., Higuchi, R.H., Horn, G.T., Mullis, K.B., and Erlich, H.A. (1988) *Science* 239:487.
14. Kinzler, K.W., and Vogelstein, B. (1989) *Nucl. Acids Res.*, in press.
15. Loh, E.Y., *et al.* (1989) *Science* 243, 217-220.
16. Frohman, M.A., Duch, M.K., and Martin, G.R. (1988) *Proc. Natl. Acad. Sci. USA* 85:8998-9002.
17. Shyamala, V., and Ames, G.F. (1989) *J. Cellular Biochem.* SBE:306.

18. Acha-Orbea, H., *et al.* (1988) *Cell* 54:263-273.
19. Urban, J.L., *et al.* (1988) *Cell* 54:577-592.
20. Lee, C.C., *et al.* (1988) *Science* 239:1288-1291.
21. Mack, D.H., and Sninsky, J.J. (1988) *Procl Natl. Acad. Sci. USA* 85:6977-6981.
22. Ludecke, H.J., Senger, G., Claussen, U., and Horsthemke, B. (1989) *Nature* 338:348-350.
23. Triglia, T., Peterson, M.G., and Kemp, D.J. (1988) *Nucl. Acids Res.* 16:81-86.
24. Ochman, H., Gerber, A.S., and Hartl, D.L. (1988) *Genetics* 120:621-623.
25. Doetschman, T., Maeda, N., and Smithies, O. (1988) *Proc. Natl. Acad. Sci. USA* 85:8583-8587.
26. Li, H., Gyllensten, U.B., Cui, X., Saiki, R.K., Erlich, H.A., and Arnheim, N. (1988) *Nature* 335:414-417.

1 : 10   1 : 50
100 : 900   70 - 980
0.33 - 24.6 - 25   1 : 250
                    4 : 996

C   vol/dilution factor
    = Times (multiplication
        action factor)
1 mg - 10⁻⁵
1 mg → mg = 1000 mg

CHAPTER 5

1000 ÷ 5 = 200

Direct Sequencing of *In Vitro* Amplified DNA

1 : 700
1000 - 5
= 995
5 : 995

Ulf Gyllensten

INTRODUCTION

The PCR method has provided a substitute for most of the repetitive types of molecular cloning and template preparation for sequencing. In combination with automated sequencing techniques, PCR will provide the fastest and most efficient means of generating nucleotide acid sequence information. The purpose of this chapter is to review methods for preparation of sequencing templates and performing direct sequencing of PCR products.

Direct sequencing has two major advantages over conventional cloning of PCR fragments into plasmids and viral genomes: 1) it can be more readily standardized (and thus amenable to automation) since it is an *in vitro* system that does not depend on living organisms (bacteria, virus), and 2) it is faster and more reliable since normally only a single sequence needs to be determined for each sample. By contrast, the sequence of several cloned PCR products have to be determined for each sample in order to distinguish mutations present in the original genomic sequence from random misincorporated nucleotides introduced by the DNA polymerase during PCR and artifacts of the amplification such as the formation of mosaic alleles ("shuffle clones") by *in vitro* recombination.

The ease with which clear and reliable sequences can be obtained directly, without resorting to cloning in bacteria, is determined by 1) the ability of the PCR primers to amplify only the target sequence (usually called the specificity of the PCR), and 2) the method used to obtain a template suitable for sequencing. The problems associated with primer specificity and the preparation of templates will be discussed below. Although chemical methods for DNA sequencing (Maxam-Gilbert) may be used for direct sequencing of PCR fragments, we will consider here only the method of Sanger[1] employing chain terminators.

OPTIMIZATION OF PCR CONDITIONS

The specificity of the PCR is to a large extent determined by the sequence of the oligonucleotides used to prime the reaction. For an individual set of primers, the specificity of the PCR can be dramatically altered by optimizing the ramp conditions, the annealing temperature, and the $MgCl_2$ concentration in the PCR buffer. A titration of $MgCl_2$ concentrations from 1.4-2.5 mM $MgCl_2$ in the final reaction in 0.2 mM increments is advised if the standard 1.5 mM concentration fails to produce the necessary specificity of the PCR.   A frequently encountered problem is that the $MgCl_2$ concentration giving the highest specificity of the PCR may sometimes result in a considerable frequency of unsuccessful PCR ("dropouts"). This may be due to relatively high concentrations of EDTA in the solution with target DNA, lowering the amount of $MgCl_2$ available for the *Taq* polymerase. For amplifications from DNA stored in TE (1 mM EDTA), it is advisable to use a PCR buffer with 2.0 mM $MgCl_2$.

The temperature ramp consists of three different stationary phases: the denaturation, annealing and extension, as well as the transition between them. For the generation of single-stranded DNA for use as a sequencing template or hybridization probe, no alteration of the cycle parameters is usually necessary.

GEL PURIFICATION OF PCR-AMPLIFIED TARGET SEQUENCES

In cases where optimization of PCR conditions fail to produce the desired specificity of the priming, either new oligonucleotides are necessary or the different PCR products can be separated by gel electrophoresis and reamplified individually for sequencing. Fragments differing in length may be separated by agarose gel electrophoresis:

1. For fragments in the size range 80-1000 bp, either a 3% NuSieve 1% regular agarose or an acrylamide gel will give sufficient resolution.

2. Cut out a thin slice of the desired PCR fragment.

3. Add 50-100 μl TE, to soak the gel slice. Either freeze and thaw the slice a few times, or leave the tube a few hours for the DNA to diffuse out of the agarose piece.

4. Take a small aliquot of the TE (1-5%) into a second PCR mix. The amount of product transferred from the eluted band to the second PCR should be less than 1 ng to obtain a clean single product.

5. Both agar and agarose have been found to contain substances that have inhibitory effects on *Taq* polymerase (M.-A. Brow, pers. comm.). Fragments that are to be reamplified for sequencing with *Taq* polymerase may preferentially be separated by acrylamide gel electrophoresis.

When the PCR primers amplify several related sequences of the same length, e.g., the same exon from several recently duplicated genes, or repetitive or conserved signal sequences, electrophoretic separation of the different products can be achieved by either of two different methods: 1) use of restriction enzymes that cut only certain templates and subsequent gel purification of the intact PCR products, and 2) the use of an electrophoretic system for separation that will differentiate between the products based on their nucleotide sequence difference. One such system, employing a denaturing formamide gradient gel[2] will be described in the next section. To be efficient, approach #1 will require some previous knowledge of the distribution of restriction sites in the different PCR products.

DIRECT SEQUENCING OF HETEROZYGOUS INDIVIDUALS

When two alleles differ by a single point mutation, direct sequencing using a PCR primer will display the heterozygote position. However, direct sequencing of allelic templates differing by several point mutations or short insertion/deletions with one of the PCR primers will generate compound sequencing ladders. There are four ways to determine the phase of several point mutations and obtain sequences of individual alleles from heterozygotes: 1) cloning apart the different templates, 2) using an electrophoretic system for separating the different templates on the basis of their nucleotide sequence prior to sequencing, 3) priming only one allele in the sequencing reaction, and 4) amplification of only one allele. An example of using the second approach is demonstrated in Figure 1. The HLA-DQα locus was amplified[3] and the alleles separated using a denaturing formamide gradient gel.[2] The DNA of the two alleles of the heterozygote individuals can then be excised from the gel, and the DNA purified and reamplified for direct sequencing of individual alleles.

The third approach is only applicable to loci where the sequence of some of the alleles is known. In the sequencing reaction, oligonucleotides made to known allele specific regions are used to selectively prime only one of the two allelic templates of a heterozygote. An example of this technique is shown in Figure 2 (panel c and d) for different alleles at the HLA-DQα

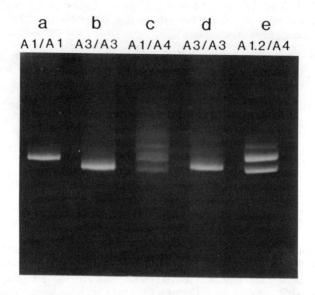

Figure 1. Separation of PCR amplified allelic variants at the HLA-DQα locus by denaturing gradient gel electrophoresis. Ten percent of the amplification reaction (100 μl) was loaded onto the top of a 20x20 cm, 12.5% polyacrylamide gel carrying a linear vertical gradient of between 25-60% denaturant.[2] The gel was submerged in buffer heated to 60°C, and run at 100 mA, 150 V, for 7 hr with recirculation of buffer. After electrophoresis, the gel was stained with ethidium bromide.

locus. The degree of selective priming achieved depends on the number and position of mismatches between the two alleles, as well as the temperature of the annealing-sequencing reaction. In the example in Figure 2, the alleles differ at multiple positions at the priming sites resulting in highly specific priming of the templates. More commonly, only a single basepair difference exists in the priming region. Unless very short primers are used, it is hard to achieve specific priming of only one allele under these conditions, since the optimal activity temperature of most thermolabile DNA polymerases (25-35°C) is too low to allow stringent annealing conditions. However, even only slightly unequal priming efficiencies may be sufficient for a discrimination between two allelic sequences. For example, allelic sequences differing by three basepairs at the 3'-end, give sufficiently unequal priming of the two alleles at room temperature (22°C) (data not shown). More specific priming can be achieved by using the thermostable *Taq* polymerase for direct sequencing, since the reaction temperature can be altered between 50-75°C.[4] This approach requires the use of a kinased sequencing primer and a single extension-termination step performed at temperatures from 50-75°C. Protocols for incorporation sequencing using *Taq* polymerase are at present not as useful for allele-specific priming since the prelabeling reaction used to produce readable sequence close to the primer has to be performed at a temperature (37-42°C) too low to significantly affect the differential priming of the two allelic templates.

The last approach is to use allele-specific oligonucleotides to amplify individual allelic templates in the PCR (Figure 2, panel g). The PCR primers can then be used for direct sequencing of individual alleles.

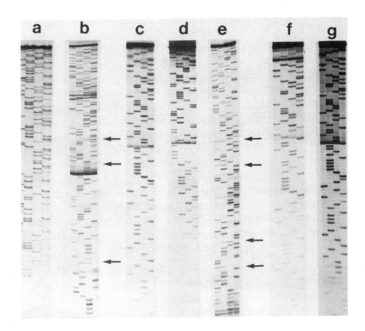

Figure 2.   Direct sequencing of single-stranded DNA from the HLA-DQα locus by the polymerase chain reaction.[3] Templates: a) KT3, a homozygous cell line sequenced with a PCR primer, b) DQA1/DQA4 heterozygous individual sequenced with a PCR primer. The two alleles differ by multiple point mutations between the PCR primers and a three basepair deletion (see arrows), c) reciprocal amplification of (b), sequenced with an oligonucleotide specific to the A3 allele, d) the same amplification as (b), sequenced with an oligonucleotide specific to the A1 allele. e) DQA4/DQA3 heterozygote sequenced with a PCR primer. The two alleles differ only by point mutations, the positions of which are indicated by the arrows, f) reciprocal amplification of (e) sequenced with an oligonucleotide for the A4 allele, g) allele specific amplification of the DQA4 allele from the DQA4/DQA3 heterozygote, sequenced with a DQA4-specific oligonucleotide.

GENERATION OF SEQUENCING TEMPLATES

Part of the problem associated with direct sequencing of PCR products is derived from the ability of the two strands of the amplified fragment to rapidly reassociate after denaturation, preventing the sequencing primer from annealing to its complementary sequence or blocking the primer-template complex from extending. Strand reassociation will permit only a fraction of the templates to participate in the sequencing reaction, resulting in weak sequencing ladders. To reduce this problem, either a variant of the standard method for sequencing double-stranded DNA may be employed or single-stranded templates may be produced in the PCR.

Double-stranded DNA Templates

Two different protocols are available for preparing templates for sequencing; both have been developed to sequence covalently closed double-stranded (dsDNA) plasmid templates. Sequencing of PCR products with these protocols is usually more difficult since the short linear templates are more prone to reassociate than the two strands of a circular plasmid. In both protocols, the PCR fragment is purified by gel electrophoresis or spin-dialysis prior to electrophoresis.

1. Denature the template in 0.2 M NaOH for 5 min at room temperature, transfer the tube to ice, neutralize the reaction by adding 0.4 volumes 5 M ammonium acetate (pH 7.5) and immediately precipitate the DNA with 4 volumes of ethanol. The DNA is resuspended in sequencing buffer and primer at the desired annealing temperature.

2. Denature the template by heat (95°C) for 5 min, quickly chill the tube by putting it on ice (or in a dry-ice ethanol bath) to slow down the reassociation of strands. Add sequencing primer and bring the reaction to the proper temperature. It is optional to add the sequencing primer after the denaturation or prior to it.

Single-stranded DNA Templates

Sequencing problems derived from strand reassociation can be avoided by preparing single-stranded DNA (ssDNA) templates either from a dsDNA template by strand separating gels or by generation of ssDNA by the PCR. Agarose strand separating gels may be successfully employed to obtain ssDNA of fragments of more than about 500 bp, they are not particularly well-suited for shorter fragments. Yet a different approach for preparing ssDNA is to use one biotinylated primer in the PCR. The two strands are then separated by passing the denatured PCR over an avidin column, that bind only the biotinylated strand. However, the simplest way is to modify the PCR in such a way that it will generate the ssDNA of choice.[3] In this procedure (asymmetric PCR), an asymmetric ratio of the two amplification primers is used to generate dsDNA for the first 20-25 cycles and, when the

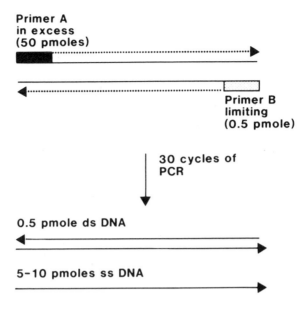

Figure 3. Outline of the procedure for generating single-stranded DNA by an asymmetric polymerase chain reaction. The concentration of the primers used in the amplification reaction are initially set to 50 pmol:0.5 pmol for a 100-μl reaction. After about 0.5 pmol of double-stranded DNA has been generated, single-stranded DNA will start to accumulate at a rate of 0.5 pmol per cycle of amplification. The resulting single-stranded DNA can be sequenced either by adding more of the limited amplification primer, or by using an internal primer.

limiting primer is exhausted, ssDNA for the next 5-10 cycles (Figure 3). Figure 4 shows the accumulation of dsDNA and ssDNA during a typical amplification of a genomic sequence (a 242-bp fragment of the HLA-DQα locus), using an initial ratio of 50 pmol of one primer to 0.5 pmol of the other primer in a 100-μl PCR. As expected, the amount of dsDNA accumulates exponentially to the point where the primer is almost exhausted, and thereafter only very slowly. In this experiment, the ssDNA generation appears to start at about cycle 25, the point where the limiting primer is almost depleted. After a short initial phase of rapid increase, the ssDNA accumulates linearly as expected when only one primer is present (primer extension). A variety of asymmetric ratios will yield ssDNA this way.[3] In general, a ratio of 50 pmol :0.5 pmol for a 100-μl PCR reaction will result in about 1-3 pmol of ssDNA after 30 cycles of PCR. The yield of ssDNA can be estimated in several ways: a) spike the PCR reaction with α^{32}P-dNTP, in addition to the normal amounts of dNTPs, run out 1% of the

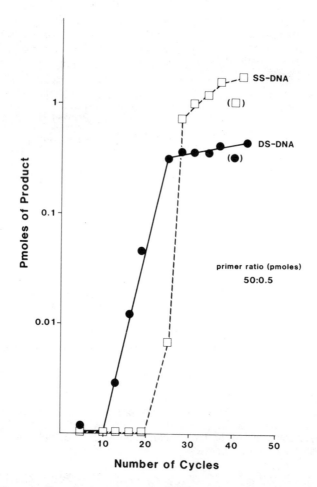

Figure 4.   Accumulation of double-stranded DNA (filled circles) and single-stranded DNA (open squares) of a 242-bp fragment of the HLA-DQα locus over 43 cycles of PCR. The primer ratio was 50 pmol:0.5 pmol in 100 µl.

reaction on a thin 3% NuSieve + 1% regular agarose gel, dry down the gel and expose it to film, b) run out 5% of the reaction, transfer it to a membrane, and probe it with an oligonucleotide complementary to the ssDNA. The ssDNA cannot be consistently quantified from staining with ethidium bromide, since the tendency of ssDNA to form secondary structures and intercalate the dye may vary between templates.

The overall efficiency of amplification appears somewhat lower (70%) when an asymmetric primer ratio is used compared to when both are present in vast excess (80-90%). In practice, this can usually be compensated for by increasing the number of PCR cycles. If the asymmetric PCR does not result in sufficient amounts of ssDNA, one of several modifications of the protocol can be tried: a) try a range of different ratios from 0.5 pmol: 50 pmol up to 5 pmol: 50 pmol, b) run 5-10 more PCR cycles, c) add more (2 units) *Taq* polymerase during the last 5 cycles, or d) try the reciprocal asymmetric primer ratio. The reciprocal asymmetric primer ratios may sometimes give different yields of ssDNA. The ssDNA generated can then be sequenced using either the PCR primer that is limiting or an internal primer and applying conventional protocols for incorporation sequencing or labelled primer sequencing. The population of ssDNA strands produced should have discrete 5' ends but may be truncated at various points close to the 3' end due to premature termination of extension. However, for any primer used in the sequencing reaction, only full-length ssDNA can be recruited as template.

Finally, an alternative approach to produce single-stranded nucleic acid templates for direct sequencing has recently been described.[5] This method involves attaching a phage promotor to one of the PCR primers, transcribing the PCR product to obtain an RNA copy and sequencing this with reverse transcriptase. This procedure has more limited applicability since it involves additional enzymatic steps following the amplification reaction, and is restricted to using reverse transcriptase as sequencing enzyme.

SEQUENCING OF PCR PRODUCTS WITH THERMOLABILE DNA POLYMERASES

A number of thermolabile DNA polymerases have been used for direct sequencing of *in vitro* amplified DNA. Detailed protocols for the use of klenow polymerase, modified T7 DNA polymerase (Sequenase), and reverse transcriptase are given at the end of the chapter (see Protocol A).

SEQUENCING OF PCR PRODUCTS WITH *Taq* POLYMERASE

Taq polymerase is an ideal enzyme for DNA sequencing, in addition to its usefulness for PCR. *Taq* polymerase has a high processivity, as well as rate of incorporation and can use nucleotide analogs such as 7-Deaza-2'-deoxyguanosine 5'-triphosphate to resolve compression of DNA sequences.[8] In addition to these properties, which it shares with T7 DNA polymerase, its high thermostability permits reaction temperatures (55-70°C) that will resolve most secondary structures. However, it has a relatively high Km for the dNTPs (roughly 10-20 μM) and an apparent absence of 3'-exonuclease activity. When dNTP concentrations drop below 1 μM, it will cause misincorporations and incorrect terminations. Agar and agarose have been found to contain inhibitors to *Taq* polymerase (A.-M. Brow, pers. comm.). These can be partly overcome by increasing the amount of *Taq* polymerase used in the sequencing reaction from 2 to 10 units. When sequencing of PCR products from cloned inserts, targets prepared by liquid lysis of phages appear to be free of the inhibitors found in targets derived from plate lysates. See Protocol B at the end of the chapter.

SEQUENCING OF REGIONS WITH STRONG SECONDARY STRUCTURE

Regions of DNA with strong secondary structure may give rise to two problems: 1) low efficiency of the PCR, due to a high frequency of templates that are not being fully extended by the *Taq* polymerase and, 2) compression of the DNA sequences in the sequencing reactions.

It appears that the high reaction temperature of PCR using *Taq* polymerase (50-75°C) should be sufficient to resolve most short secondary structures. However, strong inhibition has been observed of more complex regions, and efficient PCR of these can only be achieved after the addition of the base analog c7dGTP in the appropriate ratio relative to dGTP.[9] Similarly, base analogs may have to be used in the sequencing reactions to avoid compression problems. *Taq* polymerase will incorporate c7dGTP efficiently, but inosine should be avoided as a base analog.[9]

ERRORS INVOLVED IN SEQUENCING OF PCR AMPLIFIED DNA

The PCR can give rise to two types of discrepancies between the original target sequence and that of individual PCR products: 1) point mutational differences, and 2) *in vitro* recombined alleles. Point mutational differences due to misincorporated bases occur at a frequency of approximately 2×10^{-4} per nucleotide per cycle.[4] Over a 30-cycle amplification, individual PCR products will differ on the average every 400-4000 bp. This error rate is somewhat higher than that observed for Klenow polymerase (8×10^{-5}).[4] The *in vitro* generated alleles ("shuffle clones") are mosaic PCR products, presumably resulting from partially extended PCR products that can act as primers on other allelic templates in later cycles. Such alleles are likely to accumulate primarily in later cycles of the reaction because of insufficient enzyme to extend all available templates. Unless the two alleles in an heterozygote differ at more than one position between the PCR primers, these artifact products will never be detected.

Both types of errors have to be considered when individual PCR products are cloned and sequenced and each allelic sequence inferred from the consensus sequence of a number of clones. By contrast, "erroneous" PCR products generated by misincorporation or mosaic alleles will not interfere with the sequence determination when the PCR product is subjected to direct sequence analysis. Point mutational differences among individual PCR products will not be visible against the consensus sequence. Even if the mutation (e.g., misincorporation) arise in the very first cycle of an amplification starting from a single DNA molecule, such as when the DNA of a single sperm is amplified and sequenced, it will only appear with at the most 25% of the intensity of the consensus nucleotide at that position. For amplification reactions starting with more than a single DNA template, the frequency of PCR products containing a misincorporated nucleotide at any specific position is sufficiently low that they will not be detected. The conditions favoring the formation of mosaic alleles are less well understood, but these alleles are believed to accumulate primarily in later cycles of the reaction because of insufficient enzyme to extend all available templates. Unless very strong, secondary structure of the sequence will cause frequent termination of extension at a specific point, giving rise to artifact alleles of a single type, mosaic alleles will not confound identification of the consensus sequence.

GENERATION OF LABELED ssDNA PROBE BY PCR

As an example of the use of asymmetric PCR for generating radioactively-labeled ssDNA as hybridization probe, the central region of the second exon of the HLA-DQα locus was amplified. The oligos used are internal to a second set of PCR primers previously described for this exon.[3] One ng of a plasmid containing the HLA-DQα cDNA was used as target and 30 cycles of PCR was performed using three different molar ratios: (a) 50 pmol: 50 pmol, (b) 0.5 pmol: 50 pmol, and its reciprocal (c) 50 pmol: 0.5 pmol. To obtain radiolabeled DNA, the reaction was spiked with 5 µl α[32]P dCTP (10 µCi/µl, 1000 mCi/mmol), in addition to the cold nucleotides already present. After amplification the unincorporated nucleotides were removed using a Centricon 10 (Amicon), and 1% of the probe electrophoresed on a 4% NuSieve agarose gel, dried down, and exposed to Kodak X-Omat film. The autoradiograph shows that reaction (a) only generated labeled dsDNA, while both (b) and (c) yielded ssDNA, in addition to small amounts of dsDNA (Figure 5, a-c). The probes from reactions a-c were hybridized, without denaturation, against three nylon filters, prepared by Southern transfer of DNA from two amplifications (using 50 pmol: 0.5 pmol and 50 pmol: 0.05 pmol of the second outer set of oligonucleotides). The autoradiograph of the blot revealed that probe (a) did not hybridize (lane d) while probe (b) hybridized only to the dsDNA (lane e) and (c) to both dsDNA and ssDNA (lane f). Thus, the ssDNA produced in reaction (b) is exclusively of the desired strand and can be used as hybridization probe. To obtain labeled DNA with high specific activity, the amount of cold nucleotides have to be reduced substantially. However, due to the high Km of dNTP for *Taq* polymerase, levels below 1 µM will reduce the overall efficiency of the PCR.

AUTOMATION OF SEQUENCING REACTIONS

The repetitious nature of DNA sequencing makes it suitable for full or partial automation.[10] A schematic representation of the use of PCR in automating DNA sequencing is shown in Figure 6. A number of instruments for automated analysis of sequencing reaction are available,

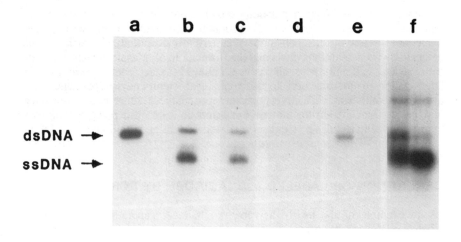

Figure 5. Generation of radioactive labeled ssDNA hybridization probe by PCR. An 82-bp fragment of the HLA-DQα locus was amplified from a cloned cDNA template using (a) 50 pmol:50 pmol, (b) 0.5 pmol:50 pmol and (c) 50 pmol:0.5 pmol of the primers, in the presence of α^{32}P-dCTP. 1% of the reactions, separated by gel electrophoresis are shown in a-c. The reactions were subsequently used to hybridize against a Southern blot filter with DNA from two previous amplifications of a region encompassing the probe area and performed with 50 pmol:0.5 pmol and 50 pmol:0.05 pmol primers.

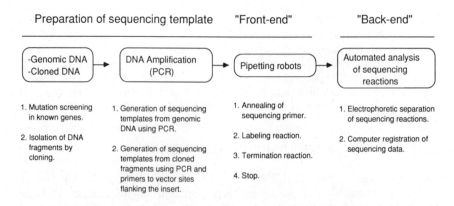

Figure 6. Use of PCR in automation of DNA sequencing.

using either conventional radiolabeled primers, or fluorescent terminators or sequencing primers. However, these instruments are only automating the "back-end" part of the analysis; the gel electrophoresis, reading and computer entering of sequences, and have to be complemented with automation of the "front-end" part of the process. Two phases of the "front-end" part can easily be automated; template preparation and sequencing reactions. PCR amplification and direct sequencing with *Taq* polymerase promises to be the most efficient method both for generation of sequencing templates and sequencing ladders. Using pipetting robots, it is possible to perform the reactions automatically, leaving only the electrophoretic analysis to be performed manually. Even laboratories that will continue to use manual electrophoretic analysis of sequencing reactions will benefit from automation of parts of the "front-end" part of the process.

ACKNOWLEDGMENTS

We would like to thank Mary Ann Brow, Henry Erlich, Russ Higuchi, Michael Innis, Randy Saiki, Stephen Scharf, and members of the laboratory of Allan C. Wilson for valuable discussion and suggestions.

PROTOCOLS

A. Direct Sequencing with Thermolabile Polymerases

1. Kinased primer sequencing with Klenow.[6]

    This is a standard protocol for kinased primer sequencing with Klenow.

    In one microfuge tube:

    -   Add 0.2-0.3 pmol double-stranded amplified DNA purified on a Centricon 30 microconcentrator or by precipitation.

    -   Add 2-2.5 pmol ^{32}P-labeled sequencing primer.

    -   The reaction is performed in 10 μl of 70 mM Tris.HCl (pH 7.6), 50 mM NaCl, 5 mM ß-mercaptoethanol, 5 mM $MgCl_2$ and 0.1 mM EDTA.

    -   Heat the mixture to 95°C for 5 min, centrifuge at 12,000 x g for 10 sec and place at 50°C.

    -   Add 5 units of Klenow (1 μl) to the primer-template mixes and immediately aliquot 2.2 μl into 2 μl of the termination mixes. Continue the reaction for 50 min at 50°C. The mixes contain: G-mix, 2 mM ddGTP, 6 μM dGTP and 60 μM of the other dNTPs; A-mix, 3.6 mM ddATP, 6 μM dATP and 60 μM of the other dNTPs; T-mix, 1.6 mM ddTTP, 6 μM TTP and 60 μM of the other dNTPs; C-mix, 1.6 mM

- Add one µl of chase solution containing 1 mM of each dNTP and continue the reaction for 15 min.

- Dry down the reaction, resuspend it in 3 µl water and add 3 µl formamide-dye stop solution (90% formamide, 20 mM EDTA, pH 8.0 and 0.05% each of the dyes xylene cyanol and bromophenol blue).

2. Kinased primer sequencing with reverse transcriptase.[7]

   In one microfuge tube:

   - Add 0.2-1.0 pmol double-tranded PCR product (purified on a Centricon 30 microconcentrator or precipitated and redissolved).

   - Add 2-4 pmol ^{32}P-labeled sequencing primer.

   - The reaction is performed in 10 µl sequencing buffer (50 mM KCl, 50 mM Tris.HCl, pH 8.0, 5 mM MgCl$_2$, 10 mM dithiothreitol).

   - Heat the reaction on ice for 10 min and then transfer the tube to ice.

   - Transfer 2.2 µl of the mixture to each of four tubes with deoxy-dideoxy mixes. The G-mix includes 20 µM dGTP, 10 µM dATP, 100 µM TTP, 100 µM dCTP and 6 µM ddGTP. The A-mix includes 100 µM dGTP, 10 µM dATP, 100 µM TTP, 100 µM dCTP, 2,5 µM ddATP. The T-mix includes 100 µM dGTP, 10 µM dATP, 20 µM TTP, 100 µM dCTP, 15 µM ddTTP. The C-mix includes 100 µM dGTP, 10 µM dATP, 100 µM TTP, 20 µM dCTP, 5 µM ddCTP. All mixes made up in sequencing buffer.

   - Add 5 units of AMV reverse transcriptase and incubate the reaction at 37°C for 15 min.

   - Add 1 µl of chase (1 mM each dNTP) and incubate for another 15 min.

   - Stop the reaction by adding 5 µl formamide-dye stop solution.

3. Incorporation sequencing with T7 DNA polymerase.[3]

   - Perform the PCR reaction as described elsewhere except for the primer amounts which are set to 50 pmol and 0.5 pmol for a 100-µl reaction. Continue the PCR for 30-35 cycles. If both strands have to be sequenced, prepare a similar reaction with the reversed primer ratio.

   - After the PCR is finished either:

     - Mix the 100 µl PCR with 2 ml of distilled water and apply to the microconcentrator Centricon 30 (Amicon) and spin at 5000 rpm in a fixed angle rotor. Dry down the retentate and resuspend in sequencing buffer (see below).

     - Precipitate the DNA in 4 M ammonium acetate to remove excess dNTPs and buffer components. Combine 100 µl PCR reaction and 100 µl 4 M ammonium acetate and mix. Add 200 µl 2-Propanol, mix, leave at room temperature for 10 min and then spin for 10 min. Remove supernatant and wash pellet carefully with 500 µl 70% Ethanol. Dry down pellet and dissolve in 10 µl TE buffer or 1x sequencing buffer (see below).

   - Use 20-60% of the PCR reaction (purified) in a total of 10 µl sequencing buffer (40 mM Tris.HCl, pH 7.5, 20 mM MgCl$_2$, 50 mM NaCl) containing

1 pmol of sequencing primer (either the limiting primer in the PCR reaction or an internal primer complementary to the ssDNA generated).

- Make the primer-template mix 65°C for 4 min and then allow it to cool down to 30°C over a period of 20 min.
- Add 1 μl of 0.1 M DTT.
- Add 2 μl of 1/20-1/100 dilution of labeling mix (1/100 dilution contains 7.5 μM of each dNTP except dATP).
- Add 0.5 μl of α^{35}S-thio dATP (1000 Ci/mmol), 10 μCi/ul.
- Add 2.0 μl of T7 DNA polymerase (diluted 1/8 in TE).
- The mixture is left at room temperature for 5 min.
- Add 3.5 μl of the labeling reaction to each of the four tubes with 2.5 μl of termination mix (each containing 80 μM of each dNTP to 8 μM of the appropriate ddNTP), and incubate the reaction at 37°C for 5 min.
- Stop the reaction by adding 4 μl formamide-dye stop solution.

B. Direct Sequencing with *Taq* Polymerase

1. Kinased primer sequencing with *Taq* polymerase.[8]

   This protocol requires no purification of the PCR product, or removal of dNTPs or buffer components before initiating the sequencing reactions. This is possible because only 20 μM of each dNTP is used in the PCR. If asymmetric PCR is used to generate the template, no denaturation is necessary.

   In one microfuge tube:

   - Add 10-17 μl (1/5) of a 100 μl PCR directly without cleanup.
     - Add 1 μl ^{32}P-labeled sequencing primer (1 pmol per μl).
   - Add 2 μl 5x buffer (5x *Taq* sequencing buffer is 35 mM MgCl$_2$, 250 mM Tris.HCl, pH 8.8) and dH$_2$O to a total of 20 μl.
   - Add 2.0 μl *Taq*-polymerase (1 unit per μl for dilution buffer, see below).
   - Dispense the reaction in four 5-μl aliquots and mix with 5 μl of the termination mixes. The termination mixes all contained 1x Buffer and 20 μM of each dNTP. In addition, the individual mixes contain: G-mix 60 μM ddGTP; A-mix, 800 μM ddATP; T-mix 800 μM ddTTP; and C-mix 400 μM ddCTP.
   - Incubate the termination reactions at 70°C for 5 min.
   - Stop the reaction by the addition of 4 μl formamide-dye stop solution.

2. Incorporation sequencing with *Taq* polymerase.[8]

   In one microfuge tube:
   - Add 7 μl DNA template (0.5 pmol).
   - Add 1 μl sequencing primer (0.5 pmol).

- Add 2 μl 5 x *Taq* sequencing buffer (5X *Taq* sequencing buffer is 250 mM Tris.HCl pH 8.8, 35 mM MgCl$_2$).

- Heat the tube to 70°C for 3 min for ssDNA templates or for 5 min at 95°C for dsDNA templates, then put the tube at 42°C for 10 min. If the reaction is performed in microtiter dishes, it may be overlayed by mineral oil to prevent evaporation.

- Add 2 μl labeling mix (dGTP or c7dGTP). The dGTP labeling mix contains 1.5 μM dGTP, 1.5 μM dCTP and 1.5 μM TTP, or for c7dGTP, 1.5 μM c7dGTP, 1.5 μM dCTP and 1.5 μM TTP.

- Add 0.5 μl α^{35}S-thio dATP (1000 Ci/mmol).

- Add 3 μl distilled water.

- Add 2 μl *Taq* polymerase (diluted from 5 units/μl to 1 unit/μl in 25 mM Tris.HCl pH 8.8, 0.1 mM EDTA, 0.15% Tween-20, 0.15% Nonidet P40).

- Mix briefly by vortexing, spin the tube, and incubate at 42°C for 5 min.

- Transfer 4 μl of the labeling reactions and mix with 2 μl of the termination mixes. The termination mixes are as described in the previous protocol (α^{32}P-labeled primer sequencing with *Taq* polymerase).

- Incubate the termination reactions at 70°C for 5 min. This time is sufficient for sequences up to at least 1000 bp in length.

- Add 4 μl formamide-dye stop solution to the termination mixes.

- Reactions may be stored at -20°C for at least a week without degradation. Heat to 70°C before loading.

REFERENCES

1. Sanger, F., Nicklen, S., and Coulson, A.R. (1979) *Proc. Natl. Acad. Sci. USA* 74:5463.
2. Fisher, S.G., and Lerman, L.S. (1983) *Proc Natl. Acad. Sci. USA* 80:1579.
3. Gyllensten, U.B., and Erlich, H.A. (1988) *Proc. Natl. Acad. Sci. USA* 85:7652.
4. Saiki, R.K., Gelfand, D.H., Stoffel, S., Scharf, S.J., Higuchi, R., Horn, G.T., Mullis, K.B., and Erlich, H.A. (1988) *Science* 239:487.
5. Stoflet, E.S., Koeberl, D.D., Sarkar, G., and Sommer, S.S. (1988) *Science* 239:491.
6. Wrischnik, L.A., Higuchi, R.G., Stoneking, M., Erlich, H.A., Arnheim, N., and Wilson, A.C. (1987) *Nucl. Acids Res.* 15:529.
7. Wong, C., Dowling, C.E., Saiki, R.K., Higuchi, R.G., Erlich, H.A., and Kazazian, Jr., H.H. (1987) *Nature* 330, 384.
8. Innis, M.A., Myambo, K.B., Gelfand, D.H., and Brow, M.A. (1988) *Proc. Natl. Acad. Sci. USA* 85:9436.
9. McConlogue, L., Brow, MA., and Innis, M.A. (1988) *Nucl. Acids Res.* 16:9869.
10. Landegren, U., Kaiser, R., Caskey, C.T., and Hood, L. (1988) *Science* 242:229.

CHAPTER 6

Using PCR to Engineer DNA

Russell Higuchi

INTRODUCTION

PCR offers a means of much more quickly and easily modifying DNA fragments to meet specific needs than was previously available. This is because of the ease of PCR itself compared to recombinant DNA techniques, because sequence changes can much more readily be made in oligonucleotide primers chemically than by manipulating DNA fragments with restriction and ligation enzymes, and because PCR products can readily accept such sequence changes as 5' "add-on" sequences to the primers. Furthermore, the efficiency at which modified product is produced is nearly 100%.

Scharf *et al.*,[1] first showed that it was quite simple to introduce restriction site sequences into DNA fragments produced by PCR merely by attaching these sequences to the 5' ends of the oligonucleotides used as primers (see Figure 1). Although these sequences are mismatched to the template DNA, in most cases they have little effect on the specificity or efficiency of the amplification; specificity is apparently imparted most

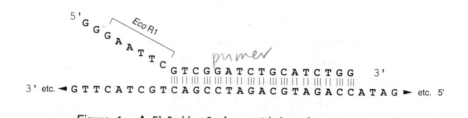

Figure 1.   A 5' "add-on" of a restriction site sequence.   A
PCR primer is shown annealed to a target sequence.   An
EcoR1 recognition sequence is shown added to the 5' end
of the primer.   Although it does not specifically match the
template sequence, this addition does not significantly affect
PCR amplification.   Thus the sequence is incorporated into
the amplification product DNA.   A few extra bases are
added 5' to the restriction site to ensure that the efficiency
of restriction enzyme cleavage is maintained.

significantly by the 3' end of the primer.   As strands initiated by these
"add-on" primers are themselves copied, the added restriction sites sequence
becomes "fixed" into the growing population of PCR product fragments.
Up to 45 base add-on sequences have been reported.[2]

This principle - the introduction of DNA alterations via the PCR primers
- is of great utility.   Most simply, it can be used to specifically label one
PCR product strand or the other or both with a radioactive, biotin, or
fluorescent tag.   It can also be used to help create DNA fragments altered
in sequence at any position in, or to recombine DNA sequences at any
desired junction.   These procedures are the subjects of this chapter.

LABELING PCR PRODUCT VIA THE PRIMERS

It is straightforward to perform a kinasing reaction on one of the two PCR
primers with γ^{32}P-labeled ATP.[3]   The labeled primer can then be substituted
directly into a standard PCR to create a product labeled at either end.
Applications of such end-labeling procedures include direct sequencing of
the product with Maxam-Gilbert procedures[4] or with phosphorothioate
chemistry[5] and the creation of DNA sequences suitable to be used in protein
binding/protection assays or "footprints."[4]   PCR is a convenient means of
creating DNA fragments for *in vitro*, protein "footprint" analysis that is
independent of restriction sites and requires fewer preparative steps than the
isolation and end-labeling of restriction fragments.

Biotin can be attached to the 5' end of an oligonucleotide primer,[6] which
remains unaffected in performance in PCR.   This has provided a means of
non-isotopic detection of hybridized PCR product[7] using streptavidin- or
avidin-enzyme conjugates, and a means of isolating one strand of product
from the other on streptavidin-columns[8] and of quantitating PCR product.[9]

Similarly, fluorescent oligos can made[10] and used in PCR. Target-specific primers with different fluors can be used to distinguish products made from different loci.[11]

ADDING USEFUL SEQUENCES TO THE ENDS OF PCR PRODUCT

As shown in Figure 1, a restriction site can be added to the 5' end of one or both oligos used in PCR. These sites facilitate insertion of the product fragments into cloning vectors. Such insertion can be "directed" by the use of different sites at each end of the product to create specific "sticky-ends." In designing add-on restriction site sequences, it seems prudent to add at least a few more bases 5' to the actual recognition sequence in case the restriction enzyme is less able to cut at an absolutely terminal sequence. Protocol A below is a procedure for cleaving PCR product at such sites and then using DNA ligase to insert the product into a cloning vector.

Other sequences that have been added to PCR fragments this way include RNA polymerase promoters to allow transcription of the amplified sequence[12] and "G+C rich clamp" sequences to DNA fragments.[2] These G+C rich sequences enhance the resolution on denaturation gradient gel electrophoresis of fragments that differ by as little as single base substitutions.

GENERALIZED MUTAGENESIS AND RECOMBINATION VIA PCR

Mullis *et al.*[13] pointed out that PCR could be used to "assemble" overlapping oligonucleotides into longer and longer PCR products in order to construct wholly synthetic DNA sequences in larger yield than could be obtained by direct chemical synthesis. The same principle, that of using PCR to combine overlapping sequences, can also be used to join precisely two PCR products made from natural DNA templates. This is shown in Figure 2, where two PCR products from different sections of the same template are made such that the resultant fragments overlap in sequence. These products can be mixed, denatured and allowed to reanneal. One of the heteroduplex forms consists of DNA strands that overlap at their 3' ends - by amplifying this heteroduplex using primers to the 5' ends of the strands a fragment can be "extracted" (to use the terminology of Mullis *et al.*[13]) from the mixture that is the precise joining of the two subsidiary PCR products. Because of the specific amplification, the desired, combined DNA fragment constitutes the great majority of PCR product.

Combined with primer-introduced sequence modification, which is limited by length constraints on the chemical synthesis of the primer, this process of PCR fragment joining is a means of introducing a sequence alteration at any position in fragment, not just at the ends. Since the overlap necessary to effect the combination need not exist in the natural template, but can be made as add-on sequences to the primers, PCR can be used to create specific site substitutions, deletions, and insertions at nearly any position

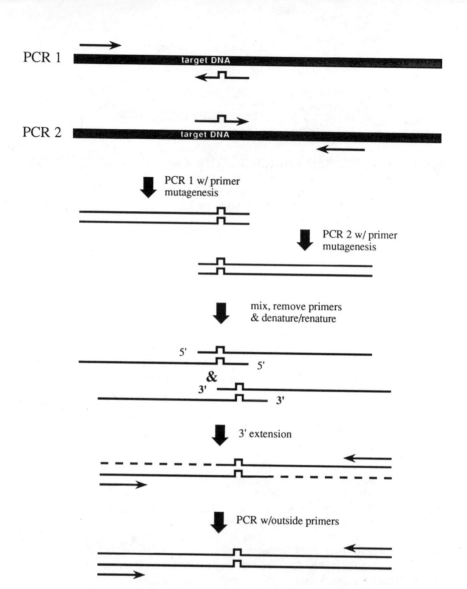

Figure 2. Site-specific mutagenesis and combining PCR fragments that overlap in sequence. Oligonucleotide primers are represented by the arrows adjacent to their annealing sites in the target DNA sequence. The two middle or "inside" primers, which anneal to the same segment of the target DNA but to the opposite strands, are shown as mismatched to the target sequence at a single base. The mismatches lead to the same sequence alteration in the two primary PCR products. PCR 1 and PCR 2 are performed separately. The products are separated from excess primers and mixed, denatured and allowed to reanneal. Some of the molecules recombine as shown through the overlap made by the middle primers. DNA chain extension on the recombinants with recessed 3' ends leads to a molecule that can be amplified with the original outside PCR primers to "extract" out a DNA fragment that has the specific base change far from its ends.

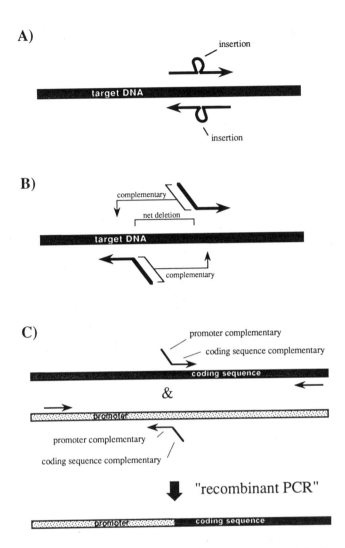

Figure 3. Insertion, deletion, and sequence recombination via PCR. A) Complementary primers that can be used to place a small insertion into the ends of two overlapping PCR fragments (the insertion is represented by the loop-out; note - the insertion sequence need not be flanked on the 5' end by template complementary sequence as shown). The insertion can be placed in the middle of a fragment by combining the two overlapping PCR fragments as in Figure 2. The size of the insert is limited by the size limitations on the primer. Longer insertion sequences can be made by making the entire insertion a PCR fragment and combining it with flanking PCR fragments as in (C) below. B) Overlapping primers that can be used to create a deletion relative to the target sequence by combining the overlapping PCR products. C) Recombinant PCR. PCR fragments from unrelated sequences can be combined specifically through complementary overlaps created by 5' add-on sequences. The overlapping PCR fragments are combined as in Figure 2.

in a DNA fragment as well as combine previously unrelated sequence at precise junctions.

These processes are illustrated in Figures 2 and 3. Figure 2 shows a specific base substitution introduced as a mismatch between a PCR primer and the target sequence. This substitution is "moved" to the middle of a DNA fragment by the use of overlapping PCR products. Figure 3a shows primers that contain as add-on sequence a small insertion. In this scenario, the size of the insertion is limited by the possible primer length. However, larger insertions can be made by making the entire insertion sequence a PCR product, and combining it with flanking sequences through overlapping primers. Figure 3b shows primers that can be used to create a deletion. Figure 3c shows how overlapping add-on sequences can facilitate the joining of hypothetical promoter and gene sequences. The use of PCR to combine different sequences has been called "recombinant PCR" (A.C. Wilson, pers. comm.). Protocol B details such combinatorial PCR.

This sort of PCR to place substitutions in the middle of a 300-800 bp PCR fragments has been demonstrated[4,14]; insertions and deletions as add-on sequences have been demonstrated by Vallette *et al.*[15]; and the precise recombining of four different sequences to create a 970-bp DNA fragment coding for a chimeric mouse class I MHC protein has been shown very elegantly by Horton *et al.*,[16] who refer to the procedure as SOE, or Splicing by Overlap Extension.

UNWANTED MUTAGENESIS IN PCR

If the PCR product is to be used in the aggregate, then low-level, random mutagenesis during the amplification should not have a significant effect as the great majority of molecules have at any one position an unaltered nucleotide.[17] However, if the PCR product is to be cloned, for example, into an expression vector, then the chance that the one molecule cloned contains a sequence alteration at other than the desired location may be significant. This may be the most serious drawback of using PCR compared to recombinant DNA means.

In a reversion assay, *Taq* DNA polymerase has been estimated to incorporate an incorrect nucleotide once every 9000 nucleotides incorporated and to cause a frameshift once every 41,000 nucleotides.[18] It would be expected that such a rate of misincorporation would result, after 20 PCR doublings of copy number, in DNA molecules with random mutations on average once every 900 bases.[19] Misincorporation under conditions of PCR, however, does not necessarily always result in sequence alterations in the final product. Misincorporated bases - which, because Taq DNA polymerase has no 3' to 5' proofreading exonuclease, cannot be removed - can promote termination of the extending DNA chain.[20] If this occurs, these errors will not propagate in subsequent PCR cycles.

PCR product that has been cloned from the same amplification and sequenced has shown a wide range of frequency of sequence change. In one study, clones of PCR amplified HLA gene sequences were compared and seen to differ on average at 1 of 400 positions after about 25 doublings. Goodenow *et al.*,[21] however, examined clones of PCR amplified HIV sequences and found that the frequency of differences in sequence between clones was at least 10-fold lower than this for a similar number of doublings. The specific mutagenesis studies of Ho *et al.*[14] and Horton *et al.*,[16] who took pains to limit the number of rounds of amplification required, found 1 unwanted mutation in 3900 bases sequenced and 1 unwanted mutation in 1800 bases sequenced, respectively - a rate that was acceptably low given the size fragments made. It's not yet clear what is chiefly responsible for these differences, but the number of doublings, conditions of amplification, as well as the sequence that is being amplified, will affect the frequency of unwanted mutagenesis.

For now, reaction conditions with dNTP concentrations between 50 and 200 micromolar are recommended (higher concentration may promote misincorporation), as are primer annealing temperatures as high as possible (so that misincorporations are more likely to lead to chain termination). The total length of time the reaction remains at high temperature should be as short as possible to prevent DNA damage that could promote more misincorporation; minimal PCR cycle times should therefore be used. On the PECI thermocycler, 30 seconds at the denaturation temperature should be adequate. Fragments of up to 1 kb can be made with as little as 30 seconds allowed for primer annealing and chain extension (R. Saiki, pers. comm.).

Since the accumulation of mutations in PCR product is proportional to the number of replications of the DNA that have occurred, the fewer PCR cycles required to provide an adequate yield the better. This argues for starting with as much starting template as possible. Note that this of course can be a trade-off as the difference between 10 and 20 doublings is a theoretical 2-fold less error frequency, but a 1000-fold difference in amplification.

Until more experience with these procedures is had, it is recommended that cloned constructs made by PCR be sequenced to verify that no unwanted mutations have occurred. Both the need to sequence and the lower likelihood of unwanted mutagenesis in smaller targets argue for as small as possible constructs, at least at first.

CONCLUSION

The use of the PCR primer, which is straightforward to chemically modify, to introduce desirable modifications into PCR product, is a recurring theme in PCR applications. It is the simplest way to add labels and/or new sequences to the ends of DNA fragments. It is also the means for creating overlapping sequences between two PCR products such that they can be easily recombined.

Recombinant PCR (to lump procedures that combine overlapping PCR fragments under one name) is potentially a powerful new tool in the study of novel macromolecules. It appears to be the most facile way yet of creating specific new DNA sequence combinations. Eventually, the need for recombinant DNA to test novel proteins coded for by these constructs may be largely circumvented as well. One can imagine that some form of *in vitro* transcription/translation could be applied directly to PCR product.[22] If the protein activity of interest can be assayed specifically and sensitively, the amount of protein produced *in vitro* may be enough to allow screening for the desired characteristics. The most satisfactory PCR construct could then be cloned and expressed in high yield.

Protocol A. Cloning PCR Products with Added Restriction Sites (Stephen J. Scharf, Cetus Corp.)

Restriction endonuclease digest

1. Mix 25 µl of completed PCR, 10 µl of the appropriate 10X restriction enzyme buffer (if two different enzymes are used simultaneously, as for directed cloning, use a buffer compatible with both, if possible - this constraint should be considered when choosing what sites to add-on in the first place), 5-10 units of the restriction enzyme(s), and distilled H$_2$0 to 100 µl.

2. Incubate at the optimal temperature for the enzyme(s) 2-3 hr.

3. Inactivate the enzyme(s) by heating to 70°C, if they are heat labile, or by extraction with an equal volume of buffered phenol[3] if they are not.

Preparation of restricted DNA for ligation

Since dNTPs carried over from the PCR may inhibit ligation, and since the DNA needs to be concentrated before ligation, purifying the DNA by either isopropanol precipitation in the presence of ammonium acetate[23] or by adsorption to and elution from "glassmilk" (Geneclean, Bio 101, La Jolla, CA) is recommended.

Ligation is carried out adding 25% of this purified DNA to 200 ng of digested vector DNA in a 10-µl reaction as per Maniatis *et al.*[3]

Transformation efficiency can sometimes be improved after ligation by extraction of the DNA with phenol and precipitation with ethanol.

Protocol B. Specific Mutagenesis and Recombination with PCR

I. Primer design. In general, primers have at least 15 bases of target sequence complementarity that are 3' to any add-on sequences. Single-base mismatches can probably be anywhere but in the very 3' 3-4 bases. Add-on sequences that provide overlap between PCR products should probably be at least 15 bases long, but the longer, probably the better, as this would stabilize the overlapping hybrid molecules.

II. "Primary" PCRs. As discussed above, to prevent unwanted mutagenesis one should begin with as much template DNA as possible. Note, however, that the relative amounts of PCR product with and without the desired mutagenesis is proportional to the amount of original template that is carried over into subsequent, combinatorial PCRs. This proportion can be decreased, if necessary, by gel purification of the primary PCR products. It's also probably a good idea to not use too many cycles of PCR past where the "plateau" in yield is reached.

III. Removal of excess primers from PCRs. In order to favor production of the combined PCR product, it is necessary to remove the overlapping or "inside" primers (see Figure 2) from the primary PCR products and then add back the "outside" primers. A physical separation of PCR product and primers is most easily accomplished by selective filtration on a Centricon 100 (ref. 4; Amicon). A protocol for this is:

1. Place 10-50 µl of the completed primary PCRs into 2 ml of TE (10 mM Tris.HCL, ph 8, 0.1 mM EDTA) in the upper reservoir of the device; cover with parafilm and invert to mix.

2. Centrifuge the device in a fixed-angle rotor for 25 min at 1000 x G. The PCR product, but not the primers, are retained. Force in excess of 1000 x G may result in some passage of product through the filter. Add another 2 ml TE to the upper reservoir and repeat the spin.

3. Recover the retentate (about 40 µl) containing PCR product as per the manufacturer's instructions.

Alternatively, products can be purified by gel-electrophoresis, which has, as noted above, the possible advantage of removing the original template DNA.

IV. Combinatorial PCR. After mixing together the primary PCR products, the creation and extension of the overlapped molecules will take place under PCR conditions. Use as large amount of the primary PCR product as is reasonable to limit DNA doublings and thereby unwanted mutagenesis. As a positive control, one should verify that the outside primers do in fact work well to amplify DNA from unmutagenized template, if this is possible (it won't be possible if one is combining previously non-contiguous sequences).

REFERENCES

1. Scharf, S.J., Horn, G.T., and Erlich, H.A. (1986) *Science* 233:1076.
2. Scheffield, V.C., Cox, D.R., Lerman, L.S., and Myers, R.M. (1989) *Proc. Natl. Acad. Sci. USA* 86:232.
3. Maniatis, T., Fritsch, E.F., and Sambrook, J. (1982) *Molecular Cloning*, Cold Spring Harbor Laboratory, p. 122.
4. Higuchi, R., Krummel, B., and Saiki, R.K. (1988) *Nucl. Acids Res.* 16:7351.
5. Nakamaye, K.L., Gish, G., Eckstein, F., and Vosberg, H.P. (1988) *Nucl. Acids Res.* 16:9947.
6. Levenson, C. (1989) in *PCR Protocols: A Guide to Methods and Applications*, Acadmeic Press, M., Innis *et al.*, eds., in press.
7. Saiki, R.K., Walsh, P.S., Levenson, C., and Erlich, H.A. (1989) *Proc. Natl. Acad. Sci. USA*, in press.

8.   Mitchell, L.G., and Merril, C.R. (1989) *Anal. Biochem.* 178:239.
9.   Syvahen, A.C., Bengstrom, M., Tenhunen, J., and Soderlund, H. (1988) *Nucl. Acids Res.* 16:11327.
10. Smith, L.M., *et al.* (1986) *Nature 321:674-679.*
11. Chehab, F.F. (1989) *J. Cellular Biochem.* Abstracts Suppl. 13E:278.
12. Stoflet, E.S., Koeberl, D.D., Sarkas, G., and Sommer, S.S. (1988) *Science* 239:491.
13. Mullis, K., Faloona, F., Scharf, S., Saiki, R., Horn, G., and Erlich, H. (1986) *Cold Spring Harbor Symp.* 51:263.
14. Ho, S.N., Hunt, H.D., Horton, R.M., Pullen, J.K., and Pease, L.R. (1989) *Gene* 77:51.
15. Vallette, F., Mege, E., Reiss, A., and Adesnik, M. (1989) *Nucl. Acids Res.* 17:723.
16. Horton, R.M., Hunt, H.D., Ho, S.N., Pullen, J.K., and Pease, L.R. (1989) *Gene* 77:61.
17. Krawczak, M., Reiss, J., Schmidtke, J., Rosler, U. (1989) *Nucl. Acids Res.* 17:2197.
18. Tindall, K.R., and Kunkel, T.A. (1988) *Biochemistry* 27:6008.
19. Saiki, R.K., Gelfand, D.H., Stoffel, S., Scharf, S.J., Higuchi, R., Horn, G.T., Mullis, K.B., and Erlich, H.A. (1988) *Science* 239:487.
20. Innis, M.A., Myambo, K.B., Gelfand, D.H., and Brow, M.A. (1988) *Proc. Natl. Acad. Sci. USA* 85:9436.
21. Goodenow, M., Huet, T., Saurin, W., Kwok, S., Sninsky, J., and Wain-Hobson, S. (1989) *J. Acquired Immune Deficiency Syndromes,* in press.
22. Bujard, H., Gentz, R., Lanser, M., Stueber, U., Muellar, M., and Ibrahimi, I. (1987) *Meth. Enzymol.* 155:415.
23. Treco, D.A. (1988) in *Current Protocols in Molecular Biology*, J. Seidman *et al.*, eds. J. Wiley, NY, Unit 2.1

CHAPTER 7

Mutation Detection by PCR, GC-Clamps, and Denaturing Gradient Gel Electrophoresis

Richard M. Myers, Val C. Sheffield, and David R. Cox

INTRODUCTION

Denaturing gradient gel electrophoresis (DGGE) allows the separation of DNA molecules differing by as little as a single base change.[1-5] The separation is based on the melting properties of DNA in solution. DNA *Tm* molecules melt in discrete segments, called melting domains, when the *temperature or denaturant concentration is raised. Melting domains vary from about 25 base pairs (bp) to several hundred bp in length, and each melts cooperatively at a distinct temperature called a Tm. Due to the considerable contribution of stacking interactions between adjacent bases on a DNA strand to double helical stability, the Tm of a melting domain is highly dependent on its nucleotide sequence. The Tms of DNA fragments differing by even very small changes, such as a single base substitution, can differ by as much as 1.5°C. In the DGGE system, DNA fragments are electrophoresed through a polyacrylamide gel that contains a linear gradient, from top to bottom, of increasing DNA denaturant concentration. As a

71

DNA fragment enters the concentration of denaturant where its lowest temperature melting domain melts (equivalent to the Tm of the domain), the molecule forms a branched structure that has a retarded mobility in the gel matrix. If the gradient conditions are chosen properly, DNA fragments differing by single base changes begin branching, and hence slowing down, at different positions in the gel, resulting in the separation of the fragments at the end of the electrophoretic run.

DGGE can be used to detect single base changes in all but the highest temperature melting domain of a DNA fragment. For example, for a DNA fragment melting in three domains, base changes in the first two domains can be detected; however, changes in the last domain are not generally detected due to the loss of sequence-dependent migration of the fragments upon complete strand dissociation. We were able to overcome this problem with cloned DNA fragments by attaching a GC-rich segment, which we call a GC-clamp, to a DNA fragment that melts in two domains.[6,7] In the absence of the GC-clamp, only those single base changes that lie in the first melting domain of this DNA fragment were separated by DGGE; the attachment of the GC-clamp to this DNA fragment allowed the separation of essentially all single base changes that lie in the second domain.

Our initial studies with GC-clamps were done by cloning mutant DNA fragments into a plasmid vector adjacent to a 300 bp sequence comprised of 80% guanosines and cytosines, and digesting the cloned DNA with restriction enzymes that released the test fragment attached to the GC-clamp. While this approach was successful, it was difficult to devise an adaption of it for directly examining genomic DNA fragments, particularly with a GC-clamp as long as 300 bp. A first step in overcoming this problem was the observation[8] and theoretical determination[6,7] that a GC-clamp as short as 30 bp is sufficient for DGGE analysis of most DNA fragments. A second development, which was based on the polymerase chain reaction (PCR),[9-11] suggested a way to attach short GC-clamps to genomic DNA fragments. This suggestion came from a report[11] that short segments of DNA, which coded for restriction enzyme cleavage sites, can be attached to the 5' ends of the oligonucleotides used to amplify DNA fragments by PCR; these "5' tails" of the oligonucleotides, while not encoded in the genomic DNA, are efficiently incorporated onto the 5' ends of the amplified DNA fragments during PCR. We recently adapted this principle to the DGGE method and showed that a 40-45 bp GC-clamp can be attached to DNA fragments amplified from human genomic DNA, and that this GC-clamp makes it possible to detect single base changes in the attached DNA fragments that were otherwise not detected by DGGE.[12] The huge amplification of the DNA fragments by PCR during this procedure also increases the sensitivity so that very small amounts of genomic DNA are required and the signals can be easily detected by ethidium bromide staining, thus obviating the need for radioactive probes. Because hybridization to a radioactive probe is not required, this method also makes it easy to examine low-, medium- and high-copy repetitive sequences, which is sometimes difficult with other probing methods such as Southern blot analysis. In this chapter, we

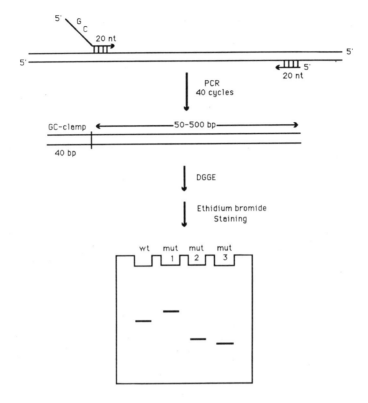

Figure 1. Scheme for attaching a GC-clamp to a DNA fragment by PCR.

describe the experimental details of this approach and discuss in particular how it can be applied towards the analysis of entire short- to medium-length genes.

STRATEGY

A depiction of the PCR/GC-clamp strategy is shown in Figure 1. Two oligodeoxyribonucleotides that flank the test segment of DNA are used to amplify genomic or cloned DNA samples by the PCR technique. One of the two oligonucleotides carries at its 5' end an additional 40-45 nt of GC-rich sequence; this 5' tail is incorporated into the 5' end of the amplified DNA fragment during PCR. The amplified DNA fragments are electrophoresed on a gradient gel containing a denaturant concentration range appropriate for the test DNA fragment, and the gel is stained with ethidium bromide and examined by UV transillumination. DNA fragments

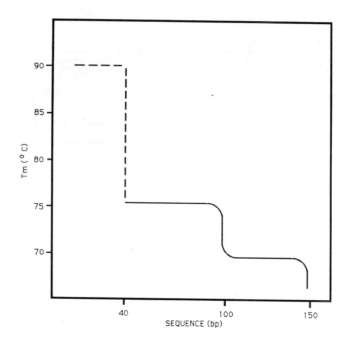

Figure 2. Attachment of a 40-bp GC-clamp to one end of a
DNA fragment by PCR allows the detection of mutations on
denaturing gradient gel electrophoresis that are otherwise
not detected. **A.** Predicted melting behavior of the mouse
β-globin promoter region.   This plot represents the melting
map of the 135-bp promoter region, which is a prediction of
the melting behavior based on its nucleotide sequence. The
x-axis represents the sequence of the promoter from -106 to
+26 relative to the start site of transcription (labeled "40"
and "150").   Along the y-axis the melting temperature
midpoint (Tm) calculated at each base pair position by the
computer algorithm of Lerman is plotted.[15]   In this map, the
DNA sequence will be melted at temperatures above the
line and remain helical at temperatures below the line.   The
fragment melts in two domains with Tms of 70°C and 75°C
of about 60 bp and 75 bp respectively.   When a 40-bp
GC-clamp is attached to the fragment at the -106 end by
PCR, an additional melting domain with a Tm above 90°C
is predicted.

differing by a single base change, which may be either a mutation or a
neutral polymorphism, will migrate to different positions in the gel and
appear as distinct ethidium-stained bands.

An example of the use of the PCR/GC-clamp approach is shown in
Figure 2. Because we had previously generated a large number of single

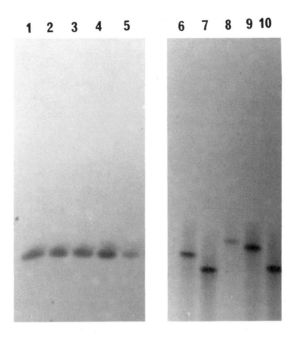

Figure 2B.   A negative image of an ethidium-stained denaturing gradient gel of wild-type and mutant DNA fragments without and with a GC-clamp attached by PCR. The wild-type mouse β-globin promoter fragment without (lane 1) and with (lane 5) a 40-bp GC-clamp. Three mutant promoter fragments without (lanes 2-4) and with (lanes 6-8) the same 40-bp GC-clamp.

base mutations in a 140-bp DNA fragment encompassing the mouse beta-major globin promoter region,[13] we used this DNA segment as a model to test the procedure. This DNA fragment was known from both theoretical predictions and experimental data to melt in two domains, the first with a Tm of 70°C and the second with a Tm of 75°C (Figure 2A).   Thus, mutations in the second (highest temperature) domain are not detectable by DGGE. According to melting theory, attachment of a GC-clamp to one end of the DNA fragment should allow mutations in the second domain to separate on the gels. Indeed, when the DNA fragment was amplified by PCR with two primers that did not incorporate a GC-clamp, none of the domain 2 mutations were detected (Figure 2B, lanes 1-5).   By contrast, attachment of a 40 bp GC-clamp to the 5' end of the promoter fragment resulted in separation of all mutations in domain 2 that were tested (Figure 2B, lanes 6-10).

The DNA fragments in Figure 2 were amplified from cloned mutant DNA templates, and are therefore equivalent to results that are obtained

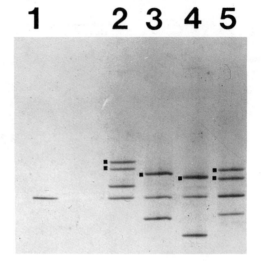

Figure 3. Heteroduplexes with a single base mismatch increase the resolution of denaturing gradient gel electrophoresis. **A.** A negative image of an ethidium-stained denaturing gradient gel that contains heteroduplex species. Lane 1 shows the position of the wild-type homoduplex fragment, which is also present in the remaining four lanes. In lanes 2-5, both homoduplexes and heteroduplexes between the wild-type DNA fragment and four different mutant fragments are present. In each lane, the black dot marks the heteroduplex species. The remaining band in each lane is the mutant homoduplex species. The mutation in lane 2 is a G to T change, causing the domain to melt at a lower temperature and therefore leads to an apparently slower mobility in the gel. In lanes 3-5, the mutations cause the domain to melt at higher temperatures, resulting in the mutant homoduplex bands running further into the gel than the wild-type homoduplex. In lanes 2 and 5, the two heteroduplexes resulting from reassorting of the strands separate from each other in the gel and two bands are seen. However, in lanes 3 and 4, the two different heteroduplexes run as a single band on this gel.

with DNA from an individual homozygous for the mutations. However, when genomic DNA is examined by this approach, most often the individual is heterozygous for the mutation or polymorphism that is detected. In these cases, the denaturing gradient gel results are more complex than the patterns seen in Figure 2. In the heterozygous individual, two homoduplex DNA fragments, comprising each of the two alleles, are present, and these two alleles appear as two separate bands in DGGE. Often, two additional bands that appear to migrate slower in the denaturing gradient gel are seen (Figure 3A). These bands are comprised of two

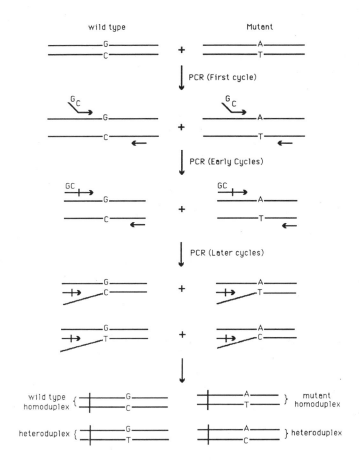

Figure 3B. Formation of heteroduplexes during PCR. In cases where an individual is heterozygous for the region of DNA being examined by PCR/DGGE, four separate species of DNA fragment are generated in later stages of the PCR procedure. This is likely due to the fact that, following denaturation of the DNA strands and during the reannealing reaction, the opposite strands of the DNA templates can outcompete or strand-displace the primer oligonucleotides because they are at such a high concentration. The end result is a reassortment of the strands into two homoduplex and two heteroduplex double-stranded DNA fragments.

different heteroduplexes, each containing a single base mismatch, formed between the two alleles during the reassorting of strands that occurs during PCR (Figure 3B). In some cases, the two heteroduplex DNA fragments do not separate from each other (Figure 3A, lanes 3 and 4 for example), resulting in a single additional band rather than two. Heteroduplexes form during the later cycles of PCR probably because the concentration of DNA builds up so high that the complementary strands reanneal and outcompete the hybridization of the oligonucleotides with their template strands. Although the pattern of three or four bands is more complex than the simple single band observed with a homozygote, it is actually advantageous to obtain the heteroduplex pattern. Because they are destabilized due to single base mismatches, heteroduplexes melt early in the denaturing gradient gel and generally separate from the wild-type band to a much greater degree than the mutant homoduplexes; in some cases, the heteroduplexes of a mutant fragment separate from the wild-type even when the homoduplex does not. Therefore, heteroduplexes lead both to increased resolution and the detection of some mutants that are not easily detected otherwise.

Several issues should be considered before beginning to use the PCR/GC-clamp procedure. As shall be discussed below, although DNA fragments as large as 1,000 bp can be examined by DGGE, the method works best with DNA fragments 500 bp and smaller. Several different strategies can be used to choose the oligonucleotides for optimum results during both the PCR amplification and DGGE. In addition, three different approaches for determining the optimum denaturing gradient and the length of time of electrophoresis for each amplified DNA fragment can be used.

OLIGONUCLEOTIDE DESIGN

In preparation for the PCR/GC-clamp technique, two oligonucleotides that will amplify the region of DNA to be tested must be chosen. Generally, the best strategy is to select segments between 100-500 bp in length for amplification. This size range was chosen because of a decrease in resolution between mutant and wild-type molecules that occurs when longer DNA fragments are examined by DGGE; in addition, DNA fragments longer than a few hundred bp travel so slowly in polyacrylamide gels that impractical electrophoresis times are required. Therefore, to examine an entire gene of several kb by the PCR/GC-clamp method, we recommend designing several pairs of opposing oligonucleotides that amplify the gene into overlapping fragments. Although it is possible to amplify multiple fragments simultaneously by PCR, care must be taken when amplifying adjacent segments to avoid production of undesirable multimer products. Therefore, for optimum results, each of these pairs should be used in separate PCR reactions. However, it is usually possible, depending on melting behavior of the DNA fragments, to mix two or more fragments after PCR amplification prior to DGGE so that multiple fragments can be analyzed in a single lane of the gel.

It is possible to predict the melting behavior of DNA fragments based on their nucleotide sequences,[6,7,14,15] and to use these predictions to select positions of the oligonucleotides that will generate DNA fragments that are best suited for DGGE. However, it is simpler and almost as effective to select the segments of interest without performing preliminary melting determinations, design oligonucleotides that will amplify these segments, and then optimize the analysis on DGGE after PCR amplification. To use this simplified approach, the following recommendations are provided:

1. Select two opposing nucleotide sequences of 20-25 nt that will serve as unique primers to generate the fragment of interest upon PCR. Ideally, the nucleotide sequences of the primers should be at least 50% G + C and should not contain indirectly repeated sequences.

2. Design one of the two PCR primers with an additional 40 nt of 100% G + C at its 5' end to serve as the GC-clamp. Melting predictions as well as our experience suggest that any random GC sequence will suffice. However, several suggestions may help prevent problems. Avoid repetitive stretches in the GC-clamp sequences that may lead to formation of secondary structures that could interfere with annealing of the primer to its template during PCR. It is also best to design the GC-clamp primer with more C residues than G residues, and to minimize the number of consecutive G's in the primer; this recommendation is made because the oligonucleotide synthesizers generally give poorer yields of primers containing more than a few adjacent G residues. As long as each PCR reaction is performed separately, it is satisfactory to use the same GC-clamp sequence for each set of primers. We have used six different GC-clamp sequences to amplify several different DNA fragments with equal success; one of those sequences is as follows:

    5' CGCCC GCCGC GCCCC GCGCC CGTCC CGCCG CCCCC GCCCC.......... 3'

    This sequence is then followed at the 3' end by 20-25 nt of unique sequence that will anneal to the target fragment region in the genomic DNA. Although we included a T residue in the middle of this GC-clamp, this is not recommended as it does cause a slight decrease in the Tm; we suggest that this or similar sequences that are 100% G + C be used for all clamps. Melting calculations based on sequence as well as our experience with six different GC-clamps indicate that a length of 40 nt is sufficient for maximum effectiveness. Therefore, the extra expense of longer clamps is not necessary. Although GC-clamps shorter than 40 nt may work when attached to some DNA fragments, they have been less effective with other fragments; therefore, we recommend against their use.

3. For most DNA fragments, only a single GC-clamp on one end is necessary. Melting predictions and some experimental evidence indicate that attachment of GC-clamps to both ends of a DNA fragment does not improve resolution of detection of base changes on DGGE. Although we have not done so, it may be desirable to amplify a DNA segment up to 1,000 bp in length with two oligonucleotides each containing a different GC-rich sequence, and then to digest the amplified fragment with a restriction enzyme that cleaves roughly in the middle of the DNA fragment to generate two fragments that each

contain a GC-clamp. One potential problem with this approach is that the two GC-primers may interfere with each other during PCR.

4. We always purify the oligonucleotide primers for PCR by preparative denaturing polyacrylamide gel electrophoresis; however, the purification may not be necessary with particular preparations of primers. Because the primers containing the GC-clamp are at least 60 nt long (40 nt of GC-clamp attached to 20 nt of unique sequence), preparations often contain shorter length species that confound the PCR results; these are most easily removed by preparative electrophoresis. Following purification, the primers are extracted with phenol, precipitated with ethanol, and resuspended in TE buffer (10 mM Tris.HCl, pH 8.0 and 1 mM EDTA) at a concentration of 10 pmol/ul.

THE POLYMERASE CHAIN REACTION

Many different conditions for amplifying DNA fragments by PCR have been reported, and the conditions that we have used are similar to those reported elsewhere.[9-11,16] The conditions that we have used are described in reference 17 and briefly as follows. When testing mammalian DNA, we have used 50-500 ng genomic DNA in a total volume of 50 µl. The buffer includes 67 mM Tris.HCl, pH 8.8, 6.7 mM $MgCl_2$, 16 mM ammonium sulfate, 10 mM β-mercaptoethanol, and 10% dimethyl sulfoxide. The reaction also includes 75 pmol of each deoxyribonucleoside triphosphate and 50 pmol of each oligonucleotide primer. After mixing the samples, one unit *Thermus aquaticus* DNA polymerase is added, and 100 µl mineral oil is layered over the top of the solution to prevent evaporation. The samples are incubated at 93°C for 1 min to denature the DNA fragments, and at a second temperature between 60°C and 72°C as described below for an additional 1-2 min. After amplification, the aqueous layer is transferred to another tube, extracted with phenol and precipitated with ethanol. The DNA products are then resuspended in 50 µl non-denaturing gel loading buffer.

Some problems have arisen in PCR amplification, and the following suggestions are given.

A. Spurious PCR amplification fragments: To use simple ethidium bromide staining to detect amplified DNA fragments on the denaturing gradient gels, it is necessary that the PCR amplification produces a single species of DNA fragment. Unfortunately, occasionally the reaction products are more complex than expected; for example, additional DNA fragments of different size than the target DNA fragment are produced, and sometimes they represent a larger fraction of the products than the target fragment. In these cases, the additional fragments interfere with the analysis of the test fragment on DGGE. We have taken several steps to prevent this problem:

1. We have found that one way to avoid spurious DNA fragments during PCR is to perform the reaction at the highest temperature

possible. It appears that a low temperature (45-55°C) annealing step in the reaction may allow the primers to amplify additional regions in genomic DNA other than the target segment. It is likely that the primers are not perfectly matched to these additional target sequences, but after an initial synthesis reaction occurs during PCR, these new products are good templates for annealing during subsequent amplification cycles. Therefore, we generally perform the annealing step at the highest temperature possible, usually between 55-65°C, and then extend at the highest temperature possible, which is usually between 60-72°C. These temperatures are determined empirically for each pair of oligonucleotides.

2. The use of longer (25-30 nt vs. 20 nt, not counting the GC-clamp sequences) oligonucleotide primers sometimes improves the specificity of the reaction, making spurious products less of a problem. The longer primers also allow a slightly higher annealing/extension temperature, which also increases specificity. In fact, the use of 30 nt primers allows the direct cycling between two temperatures in PCR (usually 70-72°C and 93°C), without the intermediate annealing step, and this may help prevent spurious bands.

3. In some cases, even when higher temperatures are used to perform the PCR, extra bands appear. It is likely that the oligonucleotides that are used in these cases anneal to similar or identical sequences in the genome, possibly to regions that represent low-, medium- or high-copy repeats. The easiest solution to this problem is to synthesize new oligonucleotides that amplify a slightly different fragment than that chosen initially. Because this is expensive, we recommend that initially only one of the two primers (the one without the GC-clamp) be changed. If the procedure is being used to test for base changes in repetitive regions of DNA, it should be possible to avoid problems by amplifying with oligonucleotides that hybridize outside of the repetitive regions.

4. If none of the above precautions prevents spurious amplification products, it may be possible to detect the specific target fragments over the background of non-specific fragments by using a radioactive probe complementary its sequence. Labeled probes can be used in several ways:

   a. The easiest approach for probing PCR-amplified products is to run the samples on DGGE as above, transfer the DNA to a nylon or nitrocellulose filter by electroblotting,[18-19] and probe the blot with a radioactively labeled DNA fragment that hybridizes to the desired sequence. Because the DNA fragment of interest has been amplified by PCR, this type of blot is very easy to do; signals can be seen in brief exposures with probes of low specific radioactivity.

b. Another way to probe PCR-amplified products is as follows. Synthesize a single-stranded probe, labeled at one of its 5' ends, by asymmetric PCR amplification[20] of the cloned target sequence. One of the oligonucleotide primers is used in excess of the other, producing mostly single-stranded product. After producing the single-stranded probe, it can be annealed to denatured PCR-amplified test DNA. We recommend that the target fragment be used in molar excess in the annealing reaction so that all of the probe is bound with the target DNA. We have used this approach with satisfactory results.[12] Note that this approach results in the formation of a heteroduplex containing a single base mismatch when the "wild-type" single-stranded DNA probe anneals to its complementary strand in the mutant DNA sample. As discussed above, the resolution of the DGGE system is greatly improved when heteroduplexes are examined due to the large degree of destabilization in the Tm of a melting domain caused by a mismatch.

c. A third way to use labeled probes is to prepare a double-stranded probe by PCR, labeled at the 5' end of only one strand, and anneal this as above for the single-stranded probe.

B. PCR-induced mutations: Another problem sometimes encountered when using PCR in conjunction with DGGE is an occasionally high mutation rate of the *Taq* DNA polymerase. Mutation rates are estimated to be about one mutation per 400 bp amplified in 40 cycles.[21-23] In practice, the mutation rate appears to vary, and it may be that the rate can be signficantly reduced by altering PCR conditions. In cases where the rate of 1/400 is seen, significant background bands on DGGE are observed; however, the specific bands are generally unmistakable in the background. Although no systematic study has determined the cause of the high *in vitro* mutation rate of *Taq* DNA polymerase, we have some evidence that the enzyme may be less faithful when used at temperatures lower than that at which *Thermus aquaticus* grows. Our results suggest that PCR reactions performed at higher temperatures produce fewer mutations, thus resulting in less background on DGGE. Therefore, we suggest that the strategy described above, in which the extension reaction in PCR is done at the highest possible temperature, be used to decrease the mutation rate.

DENATURING GRADIENT GEL ELECTROPHORESIS

Because descriptions of the DGGE system have been published, detailed protocols will not be presented here. Two separate publications[4,17] describe the equipment needed and details of preparing and running the gels. In addition, detailed discussions of ways to prepare samples, determine melting behavior, and determine optimum electrophoresis times appear in refs. 4 and 17. We suggest that these protocols be examined closely prior to purchasing or building the equipment needed for DGGE. Although detailed protocols do not appear here, a discussion of several features of the DGGE system, including the equipment used and methods for determining melting behavior, follow.

Equipment

The DGGE system requires electrophoresis of DNA fragments in a polyacrylamide gel that is maintained at a temperature that is just a few degrees below the Tm of the melting domains of the fragments being examined. For almost all naturally-occurring DNA sequences, a temperature of 60°C is satisfactory. Lerman and Fischer[1,2] devised a system that makes it easy to maintain a steady gel temperature during the electrophoretic run, and avoids problems produced by heating of the gel by flow of current. This gel system, which is that described in detail in references 4 and 17, uses a Plexiglas apparatus that holds gel plates so that both front and back surfaces are exposed. When the gel is prepared in the apparatus, an upper electrophoresis chamber in which the cathode will be placed is formed. The apparatus is then placed in an aquarium of heated electrophoresis buffer, such that this buffer reservoir forms the lower electrophoresis chamber into which the anode is placed. A peristaltic pump is used to circulate the buffer from the lower chamber into the upper chamber so that buffer is not depleted during the run. We have reported several vendors that build gel apparatuses that can be adapted for DGGE[4,17]; however, currently only one vendor (Green Mountain Lab Supply, 87 Central St., Waltham, MA 02154) provides the entire system as a package, including the aquarium, gel apparatus, gel plates, spacers, combs, circulating pump, heater and electrodes. We will also provide plans for constructing the gel apparatus upon request.

Melting Determinations

To use DGGE to maximum advantage, it is necessary to have some knowledge of the melting behavior of the test DNA fragment. This information dictates the best denaturant concentration range to use for the test fragment, as well as providing information concerning the optimum electrophoresis times. In addition, by determining melting behavior for several DNA fragments, one can design DGGE experiments that allow the analysis of multiple fragments in a single lane in a gel. Two approaches can be used to determine melting behavior of DNA fragments; one of these is empirical and one is based on a computer algorithm that predicts melting behavior from the nucleotide sequence of the fragment.

Perpendicular DGGE. The easiest empirical approach for determining melting behavior of a DNA fragment is to observe its electrophoretic behavior on a gel containing a gradient of denaturants perpendicular to the direction of electrophoresis.[13,17] Perpendicular gels can be prepared in the same apparatus used for standard, parallel DGGE. In perpendicular gels, the DNA sample is loaded across the width of the top of the gel, in a single large "well," and electrophoresed. Those fragments in the low denaturant side of the gel run far into the gel, because they do not encounter denaturant and therefore do not melt. DNA fragments on the high denaturant side of the gel barely enter the gel during electrophoresis because they melt extensively. At positions of the gel containing intermediate concentrations of denaturant, intermediate mobilities of the DNA fragment are seen. The ethidium-stained pattern of the DNA fragment in the gel after electrophoresis resembles a C_0t curve, and indeed, the concentrations of denaturant at which a steep transition in the curve occurs is equivalent to the Tm of the melting domain that the transition represents. This type of gel provides information concerning the number of melting domains in a DNA fragment as well as the Tms for each of the domains. The information can then be used to design a parallel denaturing gradient gel, which allows for multiple lane comparisons, that is optimal for the target fragment. Detailed protocols describing the use of perpendicular gels appear in refs. 13 and 17.

Prediction of Melting Behavior by Computer Algorithm. On the basis of solution studies of oligonucleotides and melting theory, Leonard Lerman has devised a computer algorithm that predicts the melting behavior, including the positions and Tms of melting domains, of a DNA fragment based on its nucleotide sequence.[14,15] A large number of experimental results with DNA fragments have confirmed that these predictions are accurate (see for example, ref. 7). The melting information gained from the algorithm can be used to calculate the denaturing gradient conditions and electrophoresis times that will result in maximum resolution in the gel of DNA fragments differing by single base changes. The use of this algorithm is described in refs. 14 and 15.

Interpreting Melting Data. The information obtained by perpendicular DGGE or the computer algorithm is used to determine the best gel conditions for each test DNA fragment. For a detailed explanation of the best way to use this information, references 13 and 17 should be consulted. However, here we describe these approaches to interpret two different types of melting behaviors.

1. One domain: A DNA fragment that melts as a single domain will melt as two domains when a GC-clamp is attached to one of its ends. This type of melting behavior is typical of many DNA fragments in the size range of 50-500 bp. Such fragments result in a single melting transition in a perpendicular denaturing gradient gel; the steep transition occurs as the first domain melts. A second transition is not seen in the gel because the GC-clamp does not melt, even at high denaturant concentrations. To select the best denaturant conditions for the parallel gel based on the perpendicular gel data, measure the horizontal position in the gel corresponding to the midpoint of the transition. This midpoint corresponds to the Tm of the domain, and by extrapolation of

the distance of the midpoint from the edges of the gel, the denaturant concentration that is equivalent to the Tm can be determined. The parallel gel conditions are chosen so that they bracket the Tm by about ten degrees C, which is equivalent to a range of 30% denaturants. For example, if the Tm of the first melting domain of a DNA fragment is 45% denaturants, then the parallel gel should contain a gradient from 30% to 60% denaturants.

If the computer algorithm is used to predict melting behavior instead of the perpendicular gel, the same logic applies. For instance, if the melting map predicts a first melting domain with a Tm of 75°C, with the GC-clamp melting above 90°C as the second domain, the parallel denaturing gradient gel should contain a gradient range equivalent to about 10°C around the Tm. Thus, for a predicted Tm of 75°C, the parallel gel should range from 70°C to 80°C. The conversion factor for Tm and percent denaturant is approximately 1°C for each 3% denaturant. Therefore, if the gel is run at 60°C, the denaturant concentration range for this fragment should be 30% (60°C + 10°C) to 60% (60°C + 20°C) denaturants.

2. Multiple domains: A DNA fragment that melts in two domains will melt in three domains when attached to a GC-clamp; such a fragment will give two mobility transitions when observed on perpendicular denaturing gradient gel electrophoresis. As before, a transition is not seen for the GC-clamp, as it does not melt under these gel conditions. In cases such as this where multiple melting domains are observed, the optimum denaturant conditions for parallel denaturing gradient gel electrophoresis can be estimated as above by measuring the midpoint of each mobility transition to determine the Tm of the domains. If two Tms are within 10% denaturant of each other, we recommend preparing a single parallel gel that spans this denaturant concentration range by 10% denaturant on either side (for example, if two domains melt at 30% and 40% denaturant, prepare a parallel gel of 20%-50% denaturant). However, if a DNA fragment contains two melting domains with Tms more than 10% denaturant apart, it is best to examine each domain separately on two different parallel denaturing gradient gels, each with the same 30% denaturant concentration range as that described above for the single domain fragments. For example, if a DNA fragment contains melting domains with Tms at 30% and 50% denaturants, the first domain should be examined on a 15% to 45% denaturant parallel gel, and the second domain should be examined on a 35% to 65% denaturant gel.

Applications

Once the melting determinations have been done for the test DNA fragments, the gels can be standardized for anaylsis of many samples. It is often possible to use a single gel to analyze more than one PCR amplified fragment; we recommend first testing a few samples of each fragment separately, and if no interference of the bands appears on the gels, they can be mixed prior to loading on subsequent gels. The interpretation of the results in denaturing gradient gels following PCR amplification is generally straightforward. If a mutation or polymorphism exists in the test fragment, a shift in mobility, either slower or faster than the wild-type fragment, is observed. A genomic DNA sample that is heterozygous within the target sequence will produce four species during PCR: two homoduplexes and two heteroduplexes. Heteroduplexes form during the later cycles of PCR as the concentration of amplified DNA fragments builds up to levels that allow efficient reannealing of two complementary DNA strands, even in the presence of high concentrations of competing oligonucleotide primers. Because they are destabilized, the heteroduplexes always appear to migrate slower than the cognate homoduplex species in DGGE. As discussed above, this feature of DGGE is advantageous because of the increase in resolution in the gel that the heteroduplexes provide.

Upon observation of shifted bands in DGGE, several courses of action can be taken depending on the needs of the researcher. In most cases, the GC-clamp approach would be used to screen a gene to detect mutations directly responsible for a disease.   In those cases where previously unidentified mutations are found, the mutation can be characterized by direct DNA sequencing of the PCR-amplified product.[24,25] Although it is possible to determine the DNA sequence of both alleles of a fragment in PCR-amplified samples from a heterozygous individual, the sequencing gel information is complicated because of the superposition of two different nucleotide sequences in a single set of lanes. One way to alleviate this problem would be to purify the mutant allele from the denaturing gradient gel, reamplify it by PCR, and sequence it directly. Indeed, we have had good results in manipulating DNA fragments purified from these gels.[13]

If the study is being done solely for genetic linkage information, the experiment can be performed on the pertinent individuals in the pedigree and all the data can be obtained from the denaturing gradient gel; no sequence information need be obtained. Similarly, if the system is being used to screen large numbers of individuals for a known mutation or set of mutations, enough data can be obtained directly from the denaturing gradient gel to adequately draw conclusions, especially if positive control cloned or genomic DNA samples containing the known mutations are included as markers.

In conclusion, we recommend the combined use of the PCR, GC-clamps, and DGGE for those cases where it is desirable to detect all possible base changes in DNA fragments up to about 500 bp in length. We find it practical to use the system with several sets of oligonucleotides to screen entire genes or regions of DNA up to about 2,500 bp, although of course more effort can be expended to screen even larger regions. This approach is particularly useful when it is desirable to screen for mutations in only the coding regions of a gene. Messenger RNA coding for the gene of interest can be synthesized into first-strand cDNA, and segments of the gene can be PCR-amplified with the appropriate primers to cover the entire gene. In those cases where a large region of DNA needs to be screened for polymorphism, but it is not important to detect every possible base change in the region, we recommend two other approaches. One involves the PCR-amplification of DNA fragments 2-5 kb in length followed by restriction digestion and DGGE,[26] and the other involves digestion of genomic DNA with frequent-cutter restriction enzymes, DGGE, and electroblot transfer to nylon membranes that are probed with radioactive probes[26] (M. Gray, personal communication).

ACKNOWLEDGMENTS

We thank the following people for their advice and contributions towards this and related projects involving denaturing gradient gel electrophoresis: Leonard Lerman, Tom Maniatis, Ezra Abrams, Kary Mullis, and members of our laboratories. This work was supported by The Wills Foundation, The Searle Scholars Program, a grant from the NIH to R.M.M. and D.R.C., and NIH Postdoctoral training grant GM07085 (V.C.S.).

REFERENCES

1. Fischer, S.G., and Lerman, L.S. (1983) *Proc. Natl. Acad. Sci. USA* 80:1579-1583.
2. Fischer, S.G., and Lerman, L.S. (1979) *Meth. Enzymol.* 68:183-191.
3. Myers, R.M., Lumelsky, N., Lerman, L.S., and Maniatis, T. (1985) *Nature* 313:495-498.
4. Myers, R.M., Maniatis, T., and Lerman, L.S. (1986) *Meth. Enzymol.* 155:501-527.
5. Myers, R.M., and Maniatis, T. *Cold Spring Harbor Symp. Quant. Biol.* 51:275-284.
6. Myers, R.M., Fischer, S.G., Maniatis, T., and Lerman, L.S. (1985) *Nucl. Acids Res.* 13:3111-3130.
7. Myers, R.M., Fischer, S.G., Lerman, L.S., and Maniatis, T. (1985) *Nucl. Acids Res.* 13:3131-3146.
8. Abrams, E., and Lerman, L.S., personal communication.
9. Saiki, R.K., Scharf, S., Faloona, F., Mullis, K.B., Horn, G.T., Erlich, H.A., and Arnheim, N. (1985) *Science* 230:1350-1354.
10. Mullis, K., Faloona, F., Scharf, S., Saiki, R., Horn, G., and Erlich, H. (1986) *Cold Spring Harbor Symp. Quant. Biol.* 51:263-273.
11. Mullis, K.B., and Faloona, F.A. (1987) *Meth. Enzymol.* 155:335-350.
12. Sheffield, V.C., Cox, D.R., Lerman, L.S., and Myers, R.M. (1989) *Proc. Natl. Acad. Sci. USA* 86:232-236.
13. Myers, R.M., Lerman, L.S., and Maniatis, T. (1985) *Science* 239:242-247.
14. Lerman, L.S., Silverstein, K., and Grinfeld, E. (1986) *Cold Spring Harbor Symp. Quant. Biol.* 51:285-297.
15. Lerman, L.S., and Silverstein, K. (1987) *Meth. Enzymol.* 155:482-501.
16. Kogan, S.C., Doherty, M., and Gitschier, J. (1987) *N. Engl. J. Med.* - 317:985-990.
17. Myers, R.M., Sheffield, V., and Cox, D.R. (1988) in *Genomic Analysis: A Practical Approach.* K. Davies, ed. IRL Press Limited, Oxford, pp. 95-139.

18. Church, G.M., and Kieffer-Higgins, S. (1988) *Science* 240:185-188.
19. Gray, M., personal communication; our results, data not shown.
20. Gyllensten, U., and Erlich, H. (1988) *Proc. Natl. Acad. Sci. USA* 85:7652-7656.
21. Saiki, R.K., Gelfand, D.H., Stoffel, S., Scharf, S.J., Higuchi, R., Horn, G.T., Mullis, K.B., and Erlich, H.A. (1988) *Science* 239:487-491.
22. Pääbo, S., and Wilson, A.C. (1988) *Nature* 334:387-388.
23. Dunning, A.M., Talmud, P., and Humphries, S. (1988) *Nucl. Acids Res.* 16: 10393.
24. Wong, C., Dowling, C.E., Saiki, R.K., Higuchi, R.G., Erlich, H.A., and Kazazian, H.H. (1987) *Nature* 330:384-386.
25. Engelke, D.R., Hoener, P.A., and Collins, F.S. (1988) *Proc. Natl. Acad. Sci. USA* 85:544-548.
26. Myers, R.M., Sheffield, V.C., and Cox, D.R. (1989) in *The Polymerase Chain Reaction*. R. Gibbs, H. Kazazian, and H. Erlich, eds.; Cold Spring Harbor Press, Cold Spring Harbor, NY, in press.

CHAPTER 8

Detection of Gene Expression

Ernest S. Kawasaki and Alice M. Wang

INTRODUCTION

The detection of gene expression can be accomplished in several ways. "Classical" methods (still important, of course) relied on the observation of biochemical or phenotypic changes in cells or organisms in order to determine the expression of a specific gene. Later, advances in macromolecular separation technology made feasible the identification and isolation of a particular gene product or protein molecule. With the advent of recombinant DNA techniques, it is now possible to detect and analyze the transcriptional product of any gene. There are several methods now in widespread use for studying specific RNA molecules. These methods include *in situ* hybridization,[1] Northern gels,[2] dot- or slot-blots,[2,3] S-1 nuclease assays,[4] and RNase protection studies.[5] In this chapter, we describe a new and powerful method to detect gene expression at the RNA level. Protocols will be outlined and discussed in detail and some applications will be briefly described. No attempt has been made to exhaustively review the literature in this fast moving area, so we wish to apologize beforehand to the numerous authors whose work may have not been cited in this chapter.

METHODS

Reagents For cDNA-PCR Reactions:

1. 10X PCR buffer: 500 mM KCl, 200 mM Tris.HCl (pH 8.4 at room temp), 25 mM $MgCl_2$ and 1 mg/ml nuclease free BSA. This solution is made by combining RNase free, autoclaved stock solutions. The BSA is not autoclaved but added to 1 mg/ml from a 10 mg/ml stock. The nuclease free BSA may be obtained from Bethesda Research Labs or other sources.

2. Deoxynucleotide Triphosphates: Neutralized, 100 mM solutions purchased from Pharmacia or PL-Biochemicals.   The dNTPs are combined to make a 10 mM stock solution of each dNTP using 10 mM Tris.HCl (pH 7.5) as diluent.

3. RNasin: Purchased from Promega Corporation at 20-40 units/µl.

4. Random hexamers: 100 pm/µl solution in TE (10 mM Tris.HCl, 1 mM EDTA, pH 8.0).   The hexamers are synthesized in-house or obtained from Pharmacia.

5. PCR Primers: The primers are usually 18-22 bases in length and dissolved in TE at 10-100 pm/µl.

6. Reverse transcriptase: Mo-MuLV obtained from Bethesda Research Labs at 200 units/µl.   Other sources and enzymes are suitable.   See Discussion.

7. *Taq* polymerase: Obtained from Perkin Elmer/Cetus at 5 units/µl.

8. Light white mineral oil: Obtained from SIGMA.

9. Chloroform: Any reagent grade and saturated with TE.

10. Microfuge tubes: Use only tubes that are specified for use in the Thermal Cycler.   This is an important point since ill-fitting tubes can result in very inefficient amplifications.

11. NuSieve and ME agarose: Obtained from FMC.

12. TEA electrophoresis buffer: 40 mM Tris.HCl, 1 mM EDTA and 5 mM sodium acetate, pH 7.5.

13. DNA Marker: 123 base pair ladder from BRL.

Note: Use autoclaved tubes and solutions wherever possible and wear gloves to prevent nuclease contamination from fingers.

Reverse Transcriptase Reaction

In a final vol of 20 µl 1X PCR buffer assemble the following: 1 mM of each dNTP, 1 unit/µl RNasin, 100 pmoles of random hexamer, 1 µg of total or cytoplasmic RNA and 100-200 units of BRL MuLV reverse transcriptase. Incubate 10 min at room temp, then 30-60 min at 42°C. To stop reaction, heat tube in 95°C water bath for 5-10 min, then quick chill on ice. Note that other manufacturers of reverse transcriptase recommend different amounts of enzyme than BRL. Follow the manufacturer's prescribed amounts. It is sometimes helpful to heat treat the RNA sample at 90°C for 5 min and quick chilling before adding an aliquot to the reaction mix. Presumably the heat treatment breaks up RNA aggregates and some secondary structure which may inhibit the priming step.

PCR Reaction

To the heat treated 20 µl reverse transcriptase reaction add 80 µl of 1X PCR buffer containing 10-50 pmoles each of upstream and downstream primer and 1-2 units of *Taq* polymerase. Layer 100 µl of the mineral oil on top of the solution to prevent evaporation during the thermal cycling. Run 20-50 PCR cycles depending on the abundance of the target. A "typical" PCR cycle which works well for amplifying lengths of 500 or more bases is: 95°C denaturation for 30 secs, 1 min cooling to 55°C, annealing of primers at 55°C for 30 secs, 30 secs heating to 72°C, extension of primers at 72°C for 30 secs, and 1 min heating to 95°C, etc. Other PCR cycle profiles work just as well, but be careful about times allotted for heating and cooling. If heating and/or cooling is too fast, the reaction solutions will not have time to equilibrate to the correct temperatures and amplification will be inefficient or nil.

Analysis of Products

After amplification, extract the reaction with 200-300 µl of TE-saturated chloroform. The mineral oil is soluble in the chloroform phase. Remove the upper, aqueous phase and use 5-10 µl for analysis in a 3% NuSieve-1% ME agarose composite gel made in TEA buffer. A 10% polyacrylamide gel in TBE buffer also works well. Use the 123-base pair ladder DNA as a convenient marker for size estimates of the products. Stain gel with ethidium bromide and photograph. Other analyses such as Southern gels or dot/slot-blots can be done at this point. Such standard methodologies are described in detail in three recent lab manuals.[6,7,8] Further manipulations such as subcloning, sequencing, etc., of the PCR product are described elsewhere in this volume and will not be discussed here.

DISCUSSION

We have not described methods for RNA isolation since several excellent reviews of protocols for RNA analysis have recently been published. For example, see appropriate chapters in references 6, 7 and 8. In general it is best to use purified RNA in the RNA/PCR reactions but as described in Chapter 4, unpurified cytoplasmic RNA can give good results.

We have found that 1 µg of total cytoplasmic RNA is more than sufficient for the amplification of rare mRNA species (one or few copies per cell). Since 1 µg is roughly equivalent to the amount present in the cytoplasm of 50-100,000 "typical" mammalian cells, the number of target molecules is usually >50,000 and should be easily amplifiable. The lowest number of target molecules detectable by this method is probably similar to the normal DNA PCR; i.e., it may be possible to detect a single RNA molecule. As evidence for this level of sensitivity, when known quantities of RNA transcripts (synthesized *in vitro* with T7 RNA polymerase) were serially diluted and tested, it was found that 10 molecules or less could be detected by this PCR method (A.M.W., unpublished results). Further, it was determined that leukemia-specific mRNA transcripts could be identified by this technique from the equivalent of less than one cell in the Philadelphia chromosome positive cell line, K562.[9] Thus, the isolation of poly A + RNA is not necessary, and the method should be sufficiently sensitive to meet the needs of most experimental situations.

The overall protocol has been simplified by use of PCR buffer in both the reverse transcriptase and PCR reactions. In our hands, the use of PCR buffer throughout works equally well or better than starting with reverse transcriptase buffer in the cDNA step followed by PCR buffer in the amplification cycles. One should note, however, that the PCR buffer may not be optimal for first stand DNA synthesis used in the formation of complementary DNA (cDNA) banks. We have not rigorously tested the different buffers for their ability to sustain the synthesis of very long DNA products.

The choice of reverse transcriptase does not seem crucial to the success of the protocol. MuLV reverse transcriptase from BRL and Boehringer-Mannheim work well, and the AMV enzyme from Boehringer-Mannheim also gives good results. Although we have not made a survey of all available enzymes, there is no reason to believe that reverse transcriptases from any reputable source would not be suitable for this method. For our own research, we have only used *Taq* polymerase supplied by Perkin Elmer/Cetus. Other manufacturers supply this enzyme but we have not tested them.

For first strand cDNA synthesis, priming may be accomplished by random hexamers, oligo(dT) or the downstream PCR primer. If one wishes to use oligo(dT), 0.1 µg per reaction is commonly employed. If the downsteam oligonucleotide is used for first strand priming, 10 to 50 pmoles is optimal. After the reverse transcriptase reaction, add the appropriate amounts of upstream and downstream primer for the PCR reaction as with the random hexamer approach. In certain instances, any one of the three priming methods can give an equivalent good, final amplification. However, the random priming method seems to give more consistent results, and usually the best overall yields of amplified target sequence[10,11] (E.S.K., unpublished results). After the first strand synthesis, removal of the RNA template by alkali or RNase treatment is not necessary. The 95°C heat treatment inactivates the reverse transcriptase and also denatures the RNA-DNA hybrids. The remaining RNA template does not seem to interfere with the PCR reaction.

It may be useful to titrate the primer oligonucleotides in the PCR reaction to find the lowest amount that can be used to give a "good" amplified product. We have found that a large excess of primers in the reaction usually results in the synthesis of more extraneous amplified products which will hinder subsequent analysis. As little as 5 pmoles of primers have been use to give a very clean and efficient amplification. However, it is best to optimize the amount for each sequence you wish to amplify. It is not necessary to use the entire cDNA reaction for PCR amplification. Aliquots can be used for PCR analysis using several different sets of primers; i.e., one cDNA reaction can be used to study several different RNA species. The cDNA reaction is usually diluted five-fold with PCR buffer. This is done to lower the dNTP concentration to 0.2 mM, a concentration that is more optimal for the *Taq* polymerase. The dNTP concentration should not exceed 0.2 mM because higher concentrations have been found to result in a higher misincorporation rate or mutation frequency for the *Taq* enzyme. There is usually no problem with amplification even if the dNTP molarity is as low as 0.05 mM.

The magnesium concentration is crucial, so care should be taken that addition of reagents does not lower the magnesium molarity to any great extent. Sometimes nucleic acids are dissolved in buffers containing 1 mM EDTA, and this can chelate out much of the magnesium. In general, try to keep the free magnesium concentration at about 2 mM in the PCR reaction.

Use the lowest number of PCR cycles that gives you the "cleanest" result. The use of more cycles than is necessary often results in the generation of many non-specific amplification products. Also, at a very high number of cycles contamination may become a problem, due to the amplification of extremely low amounts of contaminating target molecules.

The PCR primers are chosen to be around 18-22 bases in length and ~50% GC in content. If possible, the primers should be selected so that they reside in separate exons to inhibit amplification of any contaminating DNA in the RNA preparation. Even if amplification of the genomic sequence does occur, when the primers are in different exons the size difference between the mRNA and genomic product should be easily distinguishable. If the genomic structure is not known, use primers separated by 300-400 bases in the 5' coding region of the gene. Exons greater than this size are fairly rare in vertebrates[12] and, thus, the primers will have a good chance of being derived from different exons. If the gene in question has no introns, or you are studying RNA viruses, or you are investigating RNA transcription from proviral integrates, a thorough DNase treatment of the RNA will probably be necessary in order to obtain meaningful PCR results. It requires only miniscule amounts of contaminating genomic DNA to give false positives in this type of assay.

APPLICATIONS

Although this chapter is entitled detection of gene expression, it is obvious that the technique described here has far more uses than simply detection. Below, I will briefly describe several applications of this new method, and provide selected references for the readers who wish to delve into this subject in more detail.

Detection of Gene Expression

We (ref. 9; E.S.K. and A.M.W., unpublished results) and others[10,13-20] have used this method to amplify and detect a large number of different mRNAs from a variety of cell types, tissues and organs. Of course mere detection of the mRNAs is not novel. What is novel is the fact that many of the analyses were or can be carried out with RNA from 10 to 1000 cells[9,13] (A.M.W., unpublished results). This is orders of magnitude less starting material than normally required, which allows the researcher to design and execute experiments that were not feasible heretofore. As an example, investigators of hematopoiesis often use colony assays to determine the effect of growth factors or environment on the development of specific cell lineages. A question often asked is: What growth/differentiation factors are the cells within the colonies producing that might influence their own development? With the RNA/PCR methodology it is now conceivable to analyze the mRNA from hundreds of colonies for the presence of any number of growth or differentiation factors. The usual hybridization or antibody detection methods designed for this purpose would be extremely difficult of simply not possible. Another area in which this detection system will be extremely useful is in studies of transgenic animals. Frequently, it is important to know not only whether or not the transgene is expressed in the animal, but also in what cells, tissues or organs. With the increased sensitivity of the RNA/PCR method, one can test many sites from a transgenic without having to sacrifice the animal for material. This is of course desirable when one wishes to keep as many as possible of the transgenic animals intact for further studies or breeding purposes. Many, many more examples can be given, but we shall leave other possible detection schemes to the imagination of the reader.

Amplification of RNA Sequences for Diagnostic Purposes

In many cases the detection of a specific RNA molecule can be used to diagnose the existence of an infectious or genetic/cancerous disease. In the area of retroviral diseases, it is of importance to monitor the presence of retroviral RNA genomes or specific transcripts which are good indicators of an active infection. This has been done with HIV-1 patients,[21-23] and HTLV-1 and 2 and Moloney MuLV producing cell lines (E.S.K., unpublished results). The common cold virus, human rhinovirus, can also be easily detected by the RNA/PCR method.[24] Analysis of RNA transcripts from RNA as well as DNA viruses will also be useful for studying aspects of viral life cycles, such as latency, replication, etc.

In certain types of cancer, novel mRNAs are expressed. In the case of chronic myeloid leukemia (CML), and some forms of acute lymphocytic (ALL) and acute myeloid (AML) leukemia, a chimeric mRNA (BCR-ABL) is found only in the leukemic cells of the patient. Detection of this mRNA is a good diagnostic indicator of the presence of one of the forms of the disease.[9,16-19] During treatment of many neoplasias, the tumor cells become resistant to chemotherapy due to amplification of "drug-resistance" genes. Amplification at the DNA level does not necessarily lead to increased expression, but the presence of much higher amounts of the corresponding mRNA would be indicative of such a phenomenon. Increased mRNA levels have been found using the RNA/PCR method for multiple drug resistance (MDR)[10] and thymidylate synthetase (TS)[20] genes. Messenger RNA analysis for mutated RAS proto-oncogenes (see ref. 25 for review) may be of diagnostic or prognostic value in certain cancers. The use of mRNA has some advantages over the usual genomic DNA analysis. There is no interference from intron sequences, and it is possible to amplify the three H-, K- and N-RAS mRNA sequences with just one set of primers (E.S.K., unpublished results). This makes it much easier to check for mutations in codons 12, 13 and 61, which are suspect in the etiology of some cancers.

Amplification for Subcloning and Analysis of RNA Sequences

In this broad category are lumped a host of applications. One of the first uses for the RNA/PCR amplification method was to subclone a fragment of mouse ornithine transcarbamylase mRNA in order to determine the location of a deleterious point mutation.[11] Similarly, it has been used for the discovery of post-transcriptional editing of mRNA in mammalian cells,[26] studies of HLA disease associations,[27,28] analysis of human HPRT mutations,[29] studies of the relationship between autoimmunity and T-cell receptor sequences,[30] mutational analysis of human alkaline phosphatase,[31] studies of c-myc activation by woodchuck hepatitis virus,[32] the determination of alternative splicing in erythroid protein 4.1 mRNAs,[33] etc.

We have found that the RNA/PCR method is a most convenient way of obtaining cDNAs by amplification of mRNAs using primers derived from published sequences. This can be done by simply amplifying a portion of, or the entire coding region of an mRNA using primers that contain restriction enzyme sites at their 5' ends. After amplification, the PCR product can be cut with the appropriate enzymes and ligated into vectors suitable for use in expression or probe formation. With this protocol one can progress from an RNA sample to a cDNA clone modified for high expression in about a week. This is much simpler than the usual method of synthesizing/screening cDNA banks, subcloning the desired cDNA, mutagenizing the clone for expression purposes, etc. Variations on this theme are the use of degenerate primers[34,35] for amplification and cDNA clone isolation. The primers are derived from amino acid sequences, so it is possible to amplify specific RNA molecules when only minimal protein sequence is available. The PCR method can also be used to isolate cDNAs from rare mRNAs when only an internal sequence is known, using just a single gene-specific oligonucleotide primer.[36] Analysis of the amplified cDNA has been simplified by development of protocols for direct sequencing of the PCR product as described elsewhere in this volume. Including a phage (T7) promoter as part of one of the PCR primers has also facilitated sequencing by allowing production of large amounts of clone-specific RNA from the final PCR product.[37]

SUMMARY

We have presented protocols for the amplification of RNA sequences using modified PCR methodologies. The RNA/PCR technique is rapid, very sensitive and versatile. Applications of this new and powerful technique that will prove most useful center around the detection, subcloning and analysis of RNAs from very small starting samples. RNA from as few as ten molecules or from just one cell can be efficiently amplified to easily detectable levels. This new method not only greatly increases the sensitivity of detection of any RNA, but also facilitates the cDNA isolation and analysis of known and unknown RNA species. Possible refinements in the technique, such as easily quantifiable PCR reactions, will further broaden the utility of this protocol.

REFERENCES

1. Angerer, R.C., Cox, K.H., and Angerer, L.M. (1985) *Genetic Engineering* 7:43-65.
2. Thomas, P.S. (1980) *Proc. Natl. Acad. Sci. USA* 77:5201-5205.
3. White, B.A., and Bancroft, F.C. (1982) *J. Biol. Chem.* 257:8569-8572.
4. Berk, A.J., and Sharp, P.A. (1977) *Cell* 12:721-732.
5. Melton, D.A., Krieg, P.A., Rabagliati, M.R., Maniatis, T., Zinn, K., and Green, M.R. (1984) *Nucl. Acids Res.* 7:1175-1193.
6. Davis, L.G., Dibner, M.D., and Battey, J.F. (1986) *Basic Methods in Molecular Biology.* Elsevier, N.Y.
7. Ausubel, F.M., Brent, R., Kingston, R.F., Moore, D.D., Seidman, J.G., Smith, J.A., and Struhl, K. (1987) *Current Protocols in Molecular Biology.* John Wiley & Sons, N.Y.
8. Berger, S.L., and Kimmel, A.R. (1987) *Guide to Molecular Cloning Techniques. Methods in Ezymology, Vol. 152,* Academic Press, Inc., Orlando.
9. Kawasaki, E.S., Clark, S.S., Coyne, M.Y., Smith, S.D., Champlin, R., Witte, O.N., and McCormick, F.P. (1988) *Proc. Natl. Acad. Sci. USA* 85:5698-5702.

10. Noonan, K.E., and Roninson, I.B. (1988) *Nucl. Acids Res.* 16:10366.
11. Veres, G., Gibbs, R.A., Scherer, S.E., and Caskey, C.T. (1987) *Science* 237: 415-417.
12. Hawkins, J.D. (1988) *Nucl. Acids Res.* 16:9893-9908.
13. Rappolee, D.A., Mark, D., Banda, M.J., and Werb, Z. (1988) *Science* 241:708-712.
14. Chelly, J., Kaplan, J.-C., Maire, P., Gautron, S., and Kahn, A. (1988) *Nature* 333:858-860.
15. Harbarth, P., and Vosberg, H.-P. (1988) *DNA* 7:297-306.
16. Lee, M-S., Chang, K.S., Freireich, E.J., Kantarjian, H.M., Talpaz, M., Trujillo, J.M., and Stass, S.A. (1988) *Blood* 72:893-897.
17. Price, C.M., Rassool, S., Shivji, M.K.K., Gow, J., Tew, C.J., Haworth, C., Goldman, J.M., and Wiedemann, L.M. (1988) *Blood* 72:1829-1832.
18. Dobrovic, A., Trainor, J., and Morley, A.A. (1988) *Blood* 72:2063-2065.
19. Hermans, A., Selleri, L., Gow, J., and Grosveld, G.C. (1988) *Blood* 72:2066-2069.
20. Kashani-Sabet, M., Rossi, J.J., Lu, U., Ma, J.X., Chen, J., Miyachi, H., and Scanlon, K.J. (1988) *Cancer Res.* 48:5775-5778.
21. Murakawa, G.J., Zaia, J.A., Spallone, P.A., Stephens, D.A., Kaplan, B.E., Wallace, R.B., and Rossi, J.J. (1988) *DNA* 7:287-295.
22. Byrne, B.C., Li, J.J., Sninsky, J., and Poiesz, B.J. (1988) *Nucl. Acids Res.* 16:4165.
23. Hart, C., Schochetman, G., Spira, T., Lifson, A., Moore, J., Galphin, J., Sninsky, J., and Ou, C.-Y. (1988) *The Lancet* II:596-599.
24. Gama, R.E., Hughes, P.J., Bruce, C.B., and Stanway, G. (1988) *Nucl. Acids Res.* 16:9346.
25. Lacal, J.C. and Tronick, S.R. (1988) in *The Oncogene Handbook*. Reddy, E.P., Skalka, A.M., and Curran, T., eds. Elsevier Science Publishers B.V., Amsterdam, pp. 257-304.
26. Powell, L.M., Wallis, S.C., Pease, R.J., Edwards, Y.H., Knott, T.J., and Scott, J. (1987) *Cell* 50:831-840.
27. Todd, J.A., Bell, J.I., and McDevitt, H.O. (1987) *Nature* 329:599-604.
28. Sinha, A.A., Brautbar, C., Szafer, F., Friedmann, A., Tzfoni, E., Todd, J.A., Steinman, L., and McDevitt, H.O. (1988) *Science* 239:1026-1029.
29. Simpson, D., Crosby, R.M., and Skopek, T.R. (1988) *Biochem. Biophys. Res. Comm.* 151:487-492.
30. Acha-Orbea, H., Mitchell, D.J., Timmermann, L., Wraith, D.C., Tausch, G.S., Waldor, M.K., Zamvil, S.S., McDevitt, H.O., and Steinman, L. (1988) *Cell* 54:263-273.
31. Weiss, M.J., Cole, D.E., Ray, K., Whyte, M.P., Lafferty, M.A., Mulivor, R.A., and Harris, H. (1988) *Proc. Natl. Acad. Sci. USA* 85:7666-7669.
32. Hsu, T.-Y., Moroy, T., Etiemble, J., Louise, A., Trepo, C., Tiollais, P., and Buendia, M.-A. (1988) *Cell* 55:627-635.
33. Conboy, J.G., Chan, J., Mohandas, N., and Kan, Y.W. (1988) *Proc. Natl. Acad. Sci. USA* 85:9062-9065.
34. Lee, C.C., Wu, X., Gibbs, R.A., Cook, R.G., Muzny, D.M., and Caskey, C.T. (1988) *Science* 239:1288-1291.
35. Knoth, K., Roberds, S., Poteet, C., and Tamkun, M. (1988) *Nucl. Acids. Res.* 16:10932.
36. Frohman, M.A., Dush, M.K., and Martin, G.R. (1988) *Proc. Natl. Acad. Sci. USA* 85:8998-9002.
37. Sarkar, G., and Sommer, S.S. (1988) *Nucl. Acids Res.* 16:5197.

CHAPTER 9

PCR Amplification of Specific Sequences from a cDNA Library

J.-S. Tung, B.L. Daugherty, L. O'Neill,

S.W. Law, J. Han, and G.E. Mark

A NOVEL APPROACH TO RAPIDLY CLONING AND SEQUENCING RARE TRANSCRIPTS

Until recently, the most commonly used procedures in gene isolation required the establishment of a cDNA library from tissue or cellular RNAs. The gene of interest was then identified by screening the library with antibody[1] or DNA probes.[2] This approach has been very successful in cloning a great number of genes. However, construction and screening of a cDNA library is quite time-consuming, labor intensive and requires considerable protein sequence information for screening with oligonucleotide probes. The polymerase chain reaction (PCR) procedure[3,4] has enabled the specific amplification of a DNA species many million fold and has become an invaluable tool in molecular cloning and diagnosis. Most recently, the availability of *Taq* DNA polymerase has greatly simplified the PCR procedure. This enzyme has a broad temperature optimum centered around 75°C, and can survive repeated incubations at <95°C. More importantly, this enzyme is very processive and lacks 3' exonuclease activity.[5] It is, therefore, theoretically possible to clone out any gene from total RNA or DNA without going through all the steps of library construction and screening. To test this hypothesis, we selected a gene from which only limited protein sequence information was available as a model to develop a quick and simple procedure for gene isolation. We now describe the approach and methods used in the identification of an exocrine protein from the salivary glands of a South American bat.

An exocrine protein isolated from the salivary glands of a South American bat was purified by Dr. S. Gardell (MSDRL, West Point, PA) and a partial peptide sequence (GlyLeuGlyCysAspLeuMet) was the point of initiation of this project. To test the hypothesis that this gene could be cloned without the conventional screening procedures, we adopted the following strategy (Figure 1A): a) A cDNA library representing the transcripts present in bat salivary glands was constructed in lambda gt22[6] in such a way that all the cDNA inserts were flanked by SP6 and T7 polymerase promoters and thus could be amplified and directly sequenced using primers representing either promoter sequence. b) Four sets of primers, which in combination represented all of the sense codon possibilities (1024 sequence degenerations) for a portion of the sequenced protein, were synthesized. c) These primers were then paired with appropriate T7 or SP6 polymerase sequence primers to carry out a PCR amplification using total library cDNA as template. The PCR product thus generated should represent a portion of the cDNA coding for the protein of interest. Direct sequencing of this PCR product should yield enough sequence information to enable the synthesis of homologous primers for additional PCR amplification to generate DNA fragments which could then be combined to give the complete sequence of the cDNA coding for the protein.

PCR products were analyzed directly with 2% agarose gels. Figure 1B shows that with the four sets of mixed primers [primer I: 5'(CT)TX GG(ACGT) TG(CT) GA(CT) (CT)TX ATG3', each set was different in one nucleotide at position 6 from the 5' end (underlined)] and the primer representing the SP6 polymerase sequences, two products of about 450 bp and 550 bp were observed with each of the sets of mixed primers. The major PCR product (450 bp) was gel purified and its sequence was determined by the method described by Wong *et al.*[7] Clearly readable sequences (about 200 nucleotides) were generated by the mixed primers, even though it contained 256 degeneracies. Based on this sequence information, two more sets of primers [primer II (antisense) and primer III (sense)] were then made and used to pair with SP6 or T7 polymerase sequence primers, respectively, to carry out PCR amplification as before (Figure 1A). A 270-bp and a 420-bp fragment were generated employing primer II/SP6 and primer III/T7, respectively (Figure 1B). The DNA sequences of these two fragments were determined, and combined with the sequences obtained initially, a 467-bp "complete" nucleotide sequence was obtained (Figure 2). The additional sequences of the cDNA (270 bp + 420 bp) could be accounted for as a 200-bp polyA tail.

A

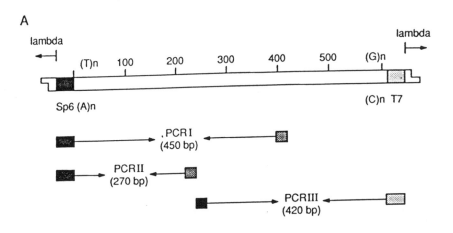

B

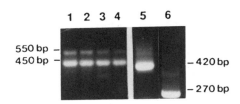

Figure 1. **A**. Strategy for PCR. (A)n, (C)n, (G)n, (T)n: homopolymer sequences SP6, T7: promoters for SP6 and T7 polymerases respectively, engineered into lambda gt vectors. **B**. Agarose gel analysis of PCR products. PCR was carried out in 100 µl volume containing 50 mM KCl, 10 mM Tris, pH 8.3, 1.5 mM $MgCl_2$, 0.01% gelatin, 200 uM of each of dNTP and 2.5 units of *Taq* polymerase (Perkin Elmer/Cetus) with 150 ng of phage library template DNA and 500 ng (0.1 nmoles) of each of the primers. The following program was used: initial template denaturation step: 8 min at 94°C, afterwards: 2 min at 94°C, 3 min at 60°C, 3 min at 72°C, for 30 cycles. The PCR reactions were extracted with chloroform to remove mineral oil. Five µl out of 100 µl was then applied to a 2% agarose gel. DNAs were visualized with ethidium bromide staining. The size of PCR products were estimated from φX-174 molecular markers (not shown). PCR products: lanes 1-4: by each of the four sets of mixed primer I and SP6 primer respectively; lane 5: by primer III and T7 primer; lane 6: by primer II and SP6 primer.

```
       10                    30                      50
       .            .        .          .            .             .
ggagtgaccacccacagcagctgcctaaccatgaagctggtgaccatcctcatgctgacc
                                 MetLysLeuValThrIleLeuMetLeuThr


       70                    90                     110
       .            .        .          .            .             .
gccctcccctgtactgctatgcggggcttggctgcgatcttatggacaatgtggtcaag
AlaLeuProLeuTyrCysTyrAlaGlyLeuGlyCysAspLeuMetAspAsnValValLys
      130                   150                    170


              .                    .                 .
ttgaccatcgattctaatgtggacgtgaatacatacatcgataacctgaaagagttccta
LeuThrIleAspSerAsnValAspValAsnThrTyrIleAspAsnLeuLysGluPheLeu
      190                   210                    230


              .                    .                 .
ccaggtgaagaaactcaagaggccttcaaattcatgaaggaatgctttctcgatcagagc
ProGlyGluGluThrGlnGluAlaPheLysPheMetLysGluCysPheLeuAspGlnSer
      250                   270                    290


              .                    .                 .
gaagaaactctggagaagatcaaagttctgcagcaatcgatatacagtagcgtttggtgt,
GluGluThrLeuGluLysIleLysValLeuGlnGlnSerIleTyrSerSerValTrpCys
      310                   330                    350


              .            .        .          .
gctcgggacactaacttccacatgaccacagagattgtccacagaccaactgtccattgg
AlaArgAspThrAsnPheHisMetThrThrGluIleValHisArgProThrValHisTrp
      370                   390                    410


              .            .        .          .
tagaagccacagctgagctttccttcttcatccttcgatctacaaatgcaagacaattgt
End
      430                   450                    470


              .            .        .          .
aaacctgacatacgtttgcatttcaataaagcatcctgcaaaaccaaaaa
```

Figure 2. Nucleotide and predicted amino acid sequences. PCR products were isolated from a preparative agarose gel and purified with Geneclean according to the manufacturer's instructions (Bio 101). 5' ^{32}P end-labeled primer was used for sequencing using a modified T7 polymerase (Sequenase) following the manufacturer's instructions (United States Biochemical Corporation Sequenase). Each reaction contained 100 ng (0.4 pmoles) of template DNA and 20 ng (4 pmoles) of 5' ^{32}P end-labeled primer representing 2×10^6 cpm per reaction. The peptide sequence that initiated this project is underlined.

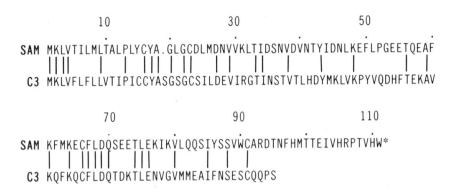

Figure 3. Sequence homologies between the bat salivary protein and the rat steroid binding protein C3 subunit precursor.

Computer aided analyses showed that this cDNA contained an open reading frame coding for 110 amino acids, with the first 18 amino acids encoding a putative signal peptide. The seven amino acid sequence used to initiate this search for the bat salivary protein was found to be located in the N-terminal region of the protein (Figure 2, underlined). A search of the NBRF data base revealed a 39% homology with the rat prostatic steroid-binding protein C3 chain precursor.[8] The most significant homology is the location of all three cysteine residues in the mature proteins (Figure 3).

We have demonstrated that polymerase chain reactions can be used to amplify a specific DNA sequence from a complex pool of molecules such as a cDNA library. In this particular experiment, the vector lambda gt22 was uniquely designed for the directional cloning of cDNA molecules. This is achieved through the use of 5' and 3' primer adapters containing appropriate restriction enzymes and polymerase promoter sequences.[6] Sequence specific probes were generated from cloned PCR products and used to screen the library by conventional means. We estimate that the abundancy of the isolated cDNA was 1 clone in every 10,000 phage recombinants. The sensitivity of this procedure is further demonstrated by the fact that specific amplification was achieved with a set of oligonucleotide primers representing a degeneracy of 256. Furthermore, as illustrated in Figure 1B, all these four mixed primers produced the same size PCR products; these results suggest that specific amplification can be achieved with oligodeoxynucleotide primer pools of even greater degeneracy and that one base-pair mismatches can be tolerated in PCR.

The combined use of PCR, mixed primers and direct sequencing of PCR products greatly reduced the time required for gene isolation and sequence verification. An additional step of reverse transcription of an RNA template[7] followed by PCR amplification can eliminate the need of establishing a cDNA library, thus extending the utility of this procedure.

ACKNOWLEDGMENTS

We thank Dr. S. Gardell for sequence information.

REFERENCES

1. Young, R.A., and Davis, R.W. (1983) *Science* 222:778.
2. Suggs, S.V., Wallace, R.B., Hirose, T., Kawashima, E.H., and Itakura, K. (1981) *Proc. Natl. Acad. Sci. USA* 78:6613.
3. Saiki, R.K., Scharf, S., Faloona, F.A., Mullis, K.B., Horn, G.T., Erlich, H.A., and Arnheim, N. (1985) *Science* 230:1350.
4. Mullis, K.B., and Faloona, F.A. (1987) *Meth. Enzymol.* 155:335.
5. Innis, M.A., Myambo, K.B., Gelfand, D.H., and Brow, M.A.D. (1988) *Proc. Natl. Acad. Sci. USA* 85:9436.
6. Han, J.H., and Rutter, W.J. (1988) in *Genetic Engineering*. Setlow, J.K., ed. Plenum Press, New York.
7. Wong, C., Dowling, C.E., Saiki, R.K., Higuchi, R.G., Erlich, H.A., and Kazazian, Jr., H.A. (1987) *Nature* 330:384.
8. Parker, M.G., White, R., Hurst, H., Needham, M., and Tilly, R. (1983) *J. Biol. Chem.* 258:12.

CHAPTER 10

Inverse Polymerase Chain Reaction

Howard Ochman, James W. Ajioka,

Dan Garza, and Daniel L. Hartl

ABSTRACT

We describe a procedure that extends the utility of the polymerase chain reaction (PCR) in allowing the geometric amplification of an unknown DNA sequence that flanks a core region of known sequence. DNA containing the core region is digested with appropriate restriction enzymes to produce a fragment of suitable size for PCR amplification. The ends of the fragment are then ligated to form a circular molecule. Primers for PCR are homologous to the ends of the core region included within the circle, but oriented such that chain elongation proceeds across the uncharacterized region of the circle rather than across the core region separating the primers. This "inverse PCR" procedure can be used to amplify the sequences that originally flanked the core sequence. Inverse PCR has applications in producing probes of anomymous sequences or in determining the sequences of upstream and downstream flanking regions themselves.

INTRODUCTION

It is often necessary to characterize a region of DNA adjacent to a segment of known sequence. Examples include the upstream and downstream flanking regions of coding DNA and the insertion sites of transposable elements as well as probes of unknown sequence from the ends of DNA fragments cloned in lambda, cosmid or yeast artificial chromosome vectors. Such end-specific probes are useful in Southern blotting or in plaque or colony hybridizations needed for chromosome walking.[1]

Probes for flanking regions are generally acquired through a series of labor-intensive and time-consuming procedures. Preliminary restriction-enzyme digestion and Southern hybridizations using probes for the known flanking region are used to identify end fragments of suitable size for cloning; and these are excised from preparative gels, cloned, and the resulting material hybridized again with the known flanking region to identify the appropriate clones. Subcloning various fragments from the clone is often necessary when the purpose is to sequence the unknown flanking region.

To bypass these procedures, we have used an extension step of the polymerase chain reaction (PCR) that allows amplification of an adjacent flanking region. Typical PCR amplifications utilize oligonucleotide primers that hybridize to opposite strands. The primers are oriented such that extension proceeds inwards across the region between the two primers. Since the product of DNA synthesis of one primer serves as the template for the other primer, the PCR procedure of repeated cycles of DNA denaturation, annealing of primers, and extension by DNA polymerase results in an exponential increase in the number of copies of the region bounded by the primers.[2-4] However, using the conventional PCR procedure, DNA sequences that lie immediately outside the primers are apparently inaccessible because oligonucleotides that prime DNA synthesis into flanking regions, rather than included regions, allow only a linear increase in the number of copies. The linear increase occurs because, for each primer, there is no priming of DNA synthesis in the reverse direction.

PCR can be used to allow the amplification of flanking regions employing a technique devised and implemented almost simultaneously by three laboratories.[5-7] The basis for this procedure ("inverse PCR") is to convert flanking DNA to interior region by cutting the molecule outside of the core region using an appropriate restriction enzyme, and forming circular molecules by self-ligation of the restriction fragments. Progressive PCR amplification of the unknown region in the circles is possible using primers homologous to the ends of the core region, but oriented with their 3' ends toward the unknown region. The procedure is outlined in Figure 1.

PROTOCOL FOR INVERSE PCR

Detailed protocols for various applications of inverse PCR can be found in Ochman *et al.*,[5] Triglia *et al.*,[6] and Silver and Keerikatte.[7] The following is a general summary of the procedure outlined by Ochman *et al.*[1]

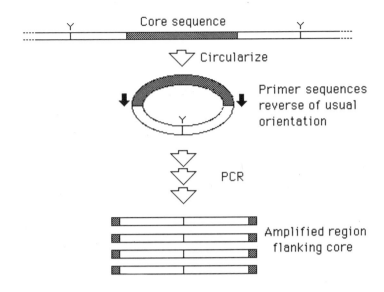

Figure 1. Amplification of flanking regions by inverse PCR.

DNA digestions are carried out using conventional buffers and other conditions recommended by the suppliers. Fragments of suitable size for inverse PCR are determined by the size of fragment that can be amplified by PCR, which at present has a practical upper limit of 3-4 kb. In many cases, preliminary Southern hybridizations will be needed to identify restriction enzymes that produce end fragments of suitable size for circularization and amplification by inverse PCR. Enzymes that cleave within the core region allow inverse PCR amplification of either the upstream or downstream segment of DNA that serves as a template for PCR primers (depending on choice of primers), whereas enzymes that do not cleave within the core region allow amplification of both flanking sequences with their junction determined by the restriction enzyme(s) and the type of circularization (e.g., ligation of complementary overhanging ends versus blunt ends). For amplification of left or right hand sequences, good choices for initial trials include enzymes with four-base recognition sites known to have conveniently located cleavage sites within the core region. If inverse PCR is to be carried out to probe hybridization probes from a large number of different sequences cloned into the same vector, it may be advisable to introduce convenient restriction sites into the vector in advance.

Circularization is performed with T4 DNA ligase in a dilute DNA concentration that favors the formation of monomeric circles.[8] In some examples, generating fragments of suitable size for inverse PCR may require the use of two restriction enzymes with ends that are incompatible for ligation, in which case the ends of the fragments should be repaired (made flush) using Klenow or bacteriophage T4 DNA polymerase prior to the circularization step. Prior to ligation, it is necessary to inactivate restriction enzymes from the previous step by phenol or heat denaturation. In our experience, it has not been necessary to cleave the circular molecules within the core region in order to obtain efficient PCR amplifications. (This was apparently not the case in the experience of Silver and Keerikatte,[7] who report an approximate 100-fold increase in efficiency of the following PCR amplification if the circle is first linearized by digestion with a restriction enzyme that cleaves within the core region. However, Triglia *et al.*[5] have found that the same effects as cleaving the circle are obtained by introducing random nicks by heating.)

Polymerase chain reaction conditions are those conventionally used.[2-9] For example, 30 cycles of denaturation at 94°C for 30 seconds, primer annealing at 58°C for 30 seconds, and extension with *Taq* polymerase at 70°C for 3 min. The PCR conditions can be altered for specific products. When inverse PCR is used in sequencing applications, it is often useful to use amplification primers set back from the ends of the core sequence, allowing the sequencing primers to be closer to the junction between the amplified part of the core sequence and the unknown flanking sequence to minimize interference from the amplification primers.[10]

APPLICATIONS OF INVERSE PCR

Generation of Flanking Regions

Applications of inverse PCR have demonstrated that the procedure can circumvent otherwise inconvenient cloning and subcloning steps, and that it can be applied to a broad range of problems. Our initial studies used inverse PCR to amplify sequences flanking the transposable insertion sequence IS*1* in natural isolates of *Escherichia coli*. Inverse PCR was applied by Triglia *et al.*[6] to the gene encoding the precursor of the major merozoite surface antigens of *Plasmodium falciparum*. In this application, they cleaved genomic DNA with *Rsa*I, ligated, the resulting circles re-cut at an internal site with *Hin*fI, and amplified. An expected 297-bp fragment was obtained and its identity confirmed by direct DNA sequencing. They suggest that inverse PCR will be useful for walking into the 5' or 3' flanking regions of transcribed genes, taking advantage of sequence information obtained from full-length cDNA.

Another application of inverse PCR was carried out by Silver and Keerikatte,[7] who used it to amplify the cellular DNA flanking integrated ecotropic proviruses in the mouse. In addition to emphasizing the use of inverse PCR in chromosome "walking" or "jumping," they point out that it could be useful in amplifying the poorly characterized sequences that are difficult or impossible to clone in *E. coli* or other host-vector systems.

Generation of End-Specific Probes

We have adapted inverse PCR in order to obtain probes specific to the insert-vector junctions in yeast artificial chromosomes (YACs).[1] The YAC library was created from high molecular weight DNA from an Oregon R strain of *Drosophila melanogaster*[11] and has an average insert size of approximately 170 kb. The Inverse PCR is applicable in this situation because it can be used to amplify a specific end of the *Drosophila* insert, using as the core region the known sequence in either arm of the YAC vector, and the resulting DNA fragment can be used as a hybridization probe to detect overlapping and adjacent clones in the library.[1]

Many *Drosophila* YAC chromosomes contain inserts so large that they hybridize with several contiguous major bands along the giant polytene salivary gland chromosomes.[11] By producing end-specific fragments from these YAC clones by inverse PCR for *in situ* hybridization, one can also determine the orientation of the insert DNA. In addition, many, if not most, YAC clones contain moderately or highly repetitive DNA sequences so they cannot be used directly for probing the library in hope of identifying overlapping clones. In Figure 2, we show an *in situ* hybridization to *Drosophila* polytene chromosomes using a biotin-labeled, end-specific probe generated by inverse PCR. This probe contains approximately 1.3 kb of *Drosophila* DNA from the 3' end of a 120-kb YAC that maps at the tip of chromosome 2R. Aside from aiding in the orientation of certain YAC clones, the end-specific DNA fragments generated by inverse PCR circumvent problems in positioning clones containing repetitive DNAs and also in the preparation of microgram quantities of probes from specific YACs for chromosome walking and *in situ* hybridization.

UTILITY AND LIMITATIONS OF INVERSE PCR

As demonstrated in Ochman *et al.*[1] and Silver and Keerikatte,[7] inverse PCR has many important applications in the study of transposable elements, retroviruses, and all other types of DNA sequences that can integrate or transpose in the genome. These applications include translocations, transpositions, or gene fusions in which one of the components is a known sequence; for example, an oncogene or genetic component of the immune system. In all these cases, a known sequence becomes inserted into or juxtaposed with unknown sequences, and inverse PCR can be used to determine the unknown flanking sequences. A major advantage of inverse PCR is simplicity and speed, allowing many independent clones to be studied. Certain applications of inverse PCR may be suitable for clinical diagnosis.

One of the present limitations derives from the unknown nature of the flanking sequence, since choice of appropriate restriction enzymes requires either pilot experiments employing several enzymes or selecting enzymes that might yield fragments of proper size. Another limitation stems from that fact that many frequently cutting restriction enzymes also cut vector sequences at unsuitable sites. However, once suitable restriction enzymes are identified, the inverse PCR procedure is straightforward and reliable.

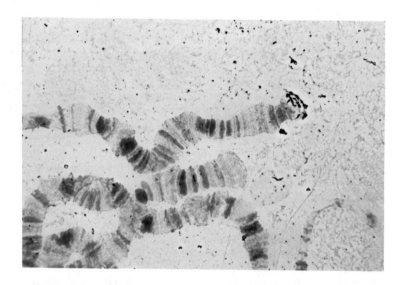

Figure 2.   *In situ* hybridization of an end-specific fragment generated by inverse PCR to polytene chromosomes of *Drosophila melanogaster*. The darkened region on the tip of chromosome 2R is the signal produced by hybridization to an inverse PCR product from a yeast artificial chromosome containing *Drosophila* genomic DNA.

Most eukaryotic genomes contain significant amounts of moderately or highly repetitive DNA, and the unknown junction sequences in YACs or cosmids will sometimes include these sequences. Probes obtained by inverse PCR amplification could potentially hybridize with many genomic sequences, and their use in chromosome walking or jumping may be limited, in which case, further subcloning would be required.

REFERENCES

1. Ochman, H., Medhora, M., Garza, D., and Hartl, D.L. (1989) in *PCR: Application & Protocols*. Innis, M., Gelfand, D., Sninsky, J., and White, T., eds. Academic Press, New York.
2. Saiki, R.K., Scharf, S., Faloona, F., Mullis, K.B., Horn, G.T., Erlich, H.A., and Arnheim, N.A. (1985) *Science* 230:1250-1354.
3. Saiki, R.K., Bugawan, T.L., Horn, G.T., Mullis, K.B., and Erlich, H.A. (1986) *Nature* 324:163-166.
4. Faloona, F., and Mullis, K.B. (1987) *Meth. Enzymol.* 155:335-350.
5. Ochman, H., Gerber, A.S., and Hartl, D.L. (1988) *Genetics* 120:621-623.
6. Triglia, T., Peterson, M.G., and Kemp, D.J. (1988) *Nucl. Acids Res.* 16:8186.

7. Silver, J., and Keerikatte, V. (1989) *J. Cell. Biochem.* Abstract #WH239, Suppl. 13E.
8. Collins, F.S., and Weissman, S.M. (1984) *Proc. Natl. Acad. Sci. USA* 81:6812-6816.
9. Saiki, R.K., Gelfand, D.H., Stoffel, S., Scharf, S.J., Higuchi, R.G., Horn, G.T., Mullis, K.B., and Erlich, H.A. (1988) *Science* 239:487-491.
10. Wrischnik, L.A., Higuchi, R.G., Stoneking, M., Erlich, H.A., Arnheim, N.A., and Wilson, A.C. (1987) *Nucl. Acids Res.* 15:529-535.
11. Garza, D., Ajioka, J.W., Burke, D.T., and Hartl, D.L. (1989) *Science*, in press.

CHAPTER 11

Alu PCR: The Use of Repeat Sequence Primers for Amplification of Human DNA from Complex Sources

David L. Nelson and C. Thomas Caskey

INTRODUCTION

The polymerase chain reaction (PCR) has revolutionized the isolation and analysis of specific nucleic acid fragments from a wide variety of sources.[1] However, application of the PCR to isolate and analyze a particular DNA region has required knowledge of DNA sequences flanking the region of interest. This limits amplification to regions of known DNA sequence. We sought to amplify human DNA of unknown sequence from complex mixtures of human and other species DNAs. In particular, we have applied the PCR to isolation of human DNA specifically from somatic cell hybrids retaining human chromosome fragments in rodent cell backgrounds. This allows the isolation and characterization of sequences from specific human regions retained in hybrids, obviating the requirement of cloned DNA libraries and isolation of human clones through the use of human-specific repeat sequence probes.[2] The method, *Alu* PCR, has also proven useful for the rapid isolation of human insert DNA from cloned sources as well, extending the application of the PCR to genomic DNAs cloned in lambda and yeast artificial chromosome (YAC)[3] vectors. The adaptation of the PCR to large genomic regions provides another tool for the analysis of the human genome.

The *Alu* PCR method makes use of the ubiquitous *Alu* repeat sequence found in human DNA. Approximately 900,000 copies of this 300-bp sequence are distributed throughout the human genome.[4] Although there is considerable variation between copies of the *Alu* repeat, a consensus sequence has been established, and there are regions of the repeat that are reasonably well conserved.[5] We reasoned that appropriate PCR primers designed to recognize these conserved regions should allow sufficient inter-*Alu* amplification for the isolation of human DNA from complex sources. In order to amplify human DNA in somatic cell hybrids without also amplifying rodent *Alu*-equivalents, we also required identification of human specific primers.

HUMAN SPECIFIC AMPLIFICATION FROM HYBRID CELL LINES

A number of primers were synthesized and tested for their ability to amplify human, somatic cell hybrid and rodent DNAs. Figure 1 shows amplification using two, TC-65 and #278. The #278 primer recognized rodent DNA and amplified rodent sequences, while the TC-65 primer was specific to human DNA and amplified a discrete number of bands from the X3000-11.1 hybrid cell line (which contains Xq24-qter as its only human material[6]). It is noteworthy that the location of the TC-65 primer is in the 31-bp primate-specific insert in the second repeat of the human *Alu* consensus sequence, while the #278 primer is located in a portion of the repeat that is highly conserved among mammalian *Alu*-equivalents.[7] Detailed description of the reaction conditions and primers involved in these reactions are contained in a manuscript submitted for publication.[8]

The following is a brief summary of our findings. The amount of sequence amplified appears dependent upon the particular oligonucleotide and the amount of target human DNA as a proportion of the total target DNA. The original effective primer, TC-65, contains 17 bases of *Alu* sequence and a 13 base 5' extension containing a *Not*I site for cloning the amplified products. TC-65 used alone allows amplification only between *Alu* sequences in opposite orientation and within a distance of up to 5-6 kb. It does not amplify every such pair of *Alu* sequences, demonstrating some preference for particular *Alu* repeats. This can be shown in the hybrid cell line X3000-11.1 (Figure 1), where amplification with TC-65 shows individual bands numbering between 50 and 100. Since this hybrid contains about 40 Mb of human DNA, and *Alu* sequences are thought to occur every four kb on average, there should be approximately 10,000 *Alu* repeats in this region. Of these, a conservative estimate of copies in opposite

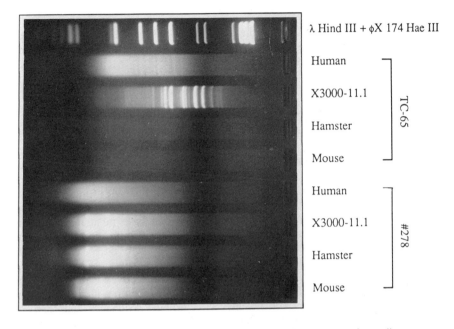

Figure 1. Amplification of human DNA from somatic cell hybrids using *Alu* primers. Results of PCR using the TC-65 and #278 primers for amplification of DNA from human, X3000-11.1, hamster and mouse. Lanes are marked for primer and DNA source. Size markers are a mix of λ DNA cut with *Hin*dIII and φX 174 DNA cut with *Hae*III. PCR was carried out in a total volume of 100 ml with 1 mg of DNA, primer at 1 mM in 50 mM KCl, 10 mM Tris pH 8.0, 1.5 mM MgCl$_2$, 0.01% gelatin, dATP, dCTP, dGTP, and dTTP at 300 mM each (Pharmacia), and 2.5 units of *Taq* polymerase (Perkin Elmer/Cetus) for 35 cycles of 94°C denaturation (1 min), 55°C annealing (45 sec), and 68°C extension (5 min) in an automated Thermal Cycler (Perkin-Elmer Cetus Instruments). Gel is 1.1% agarose (SeaKem LE-FMC Corp.) run in Tris-borate buffer.

orientation and within an amplifiable distance would be 1000. Since at most only 100 species appear with this primer, there must be characteristics of this 10% of pairs of *Alu* repeats that favor amplification with this primer.

It was with great interest that we found that one of the sequences (a 3.5-kb band) amplified by this primer from the X3000-11.1 cell line (and all other hybrid lines containing the human *hprt* locus) hybridizes with a cDNA probe for the *hprt* mRNA.[9] While confirming that a *bona fide* human-specific sequence is amplified, this sequence also affords the opportunity to investigate the sequence of the *Alu* repeats involved in this amplification. To localize the repeats within the *hprt* gene, we amplified DNA from six bacteriophage clones encompassing the entire coding region[10,11] using the same TC-65 primer.   Two of the clones demonstrate the expected 3.5-kb band, which also hybridizes to the cDNA probe. Since these clones overlap and contain only exons 7, 8 and 9, we predicted that the *Alu* repeats priming synthesis of this sequence are located 3.5 kb from one another, in opposite orientation and spanning one or more of the last three exons in the gene.   Inspection of the sequence from this region (which has been determined in a collaborative arrangement with W. Ansorge at EMBL/Heidelberg, Edwards *et al.* in preparation) has allowed us to identify the *Alu* sequences involved. The *Alu* repeats span exon 9, with one repeat located in the intron between exons 8 and 9, and the other downstream of the coding region.   The downstream *Alu* repeat involved in this amplification is a perfect match for the 17 bases of *Alu* present in the primer, but shows no particular homology to the 5' tail of the primer. The *Alu* repeat between exons 8 and 9 contains a one base mismatch with the 17 bases of *Alu* in the primer, but this may be compensated by the presence of two additional matches to the tail portion immediately adjacent to the *Alu* region, giving 18 of 19 homologous bases in a row.   This sequence information should help us understand the selective nature of the amplification with the TC-65 primer, especially since the identification of this pair of amplifiable *Alu* repeats will provide sequences that can be amplified with other primers to help determine the parameters important in this reaction.

Additional primers complementary to the same region of the *Alu* repeat have also been found to amplify human DNA specifically from hybrid cells. The reverse complement of the 17 base region in TC-65 is also useful, allowing a separate set of fragments to be amplified from human regions in hybrid cells. Primers without tails have been particularly useful, as they allow considerably more sequence to be amplified from a given DNA source.   As an example, the X3000-11.1 line when amplified by these primers shows a smear of bands numbering in the hundreds, while hybrids containing smaller amounts of human sequence (1-2 Mb) show a discrete number of bands (10-20). These primers should allow the extension of the *Alu* PCR technique to a number of different areas, including the use of amplified product from hybrids as hybridization probes to identify cloned DNAs from specific genomic regions in total human libraries and amplification from pulsed-field gel fragments. Characterization of additional oligonucleotides is vital for defining the parameters important in these reactions. We hope to identify primers that will allow amplification in both directions from a single *Alu* repeat, thus extending our ability to amplify to the majority of repeats in a region. This could be combined with "inverted" PCR (where the template is cut and re-ligated under circularizing conditions prior to PCR "around the circle"[12]), removing the requirement of two *Alu* repeats in close proximity for amplification.

Alu PCR FROM CLONED GENOMIC DNA

When combined with primers directed to the ends of vector sequences adjacent to cloning sites, the *Alu* PCR primers can allow direct isolation of human inserts in cloned DNAs. This has been particularly useful for the isolation of inserts in YAC clones, where the alternative methods are tedious and time consuming. Using this approach, we have generated fragments from about 75% of YAC clones containing human inserts, and have been able to use these fragments as probes to determine the chromosomal locations of the YAC inserts in somatic cell hybrid mapping panels.[8] This approach applied to clones in lambda phage vectors offers the opportunity to derive probes from the insert without growing the phage for DNA isolation. All lambda clones have yielded product in these reactions, and all products have given the expected result when used as probes for mapping chromosomal positions. This can save considerable time in localizing cloned DNAs. It is important to note that in contrast to amplification from somatic cell hybrids, the primers used in amplification from cloned human DNAs need not be human specific, and most *Alu* primers tested have yielded products.

The pattern of bands generated by *Alu* PCR can also be useful for analysis of human regions in somatic cell hybrids and cloned sequences. The *Alu* PCR fragments (APF) can be used as a "fingerprinting" method for determining overlaps between regions in hybrids or clones. This should prove extremely useful for the determination of the extent and nature of human sequences in somatic cell hybrids with small (1-10 Mbp) human regions, where analysis is otherwise difficult. APF may also prove to be the method of choice for determining overlaps between clones in YAC vectors, where other "fingerprinting" methods will be difficult.

Alu PCR has tremendous potential for isolation and analysis of small regions of human DNA in complex backgrounds. It may also provide an opportunity to target and isolate important sequences from similarly complex regions. It is rapid, inexpensive and easy to perform. It should become an important method for reducing the complexity of the genome.

ACKNOWLEDGMENTS

We wish to acknowledge the technical assistance of Maureen Victoria and Ramiro Ramirez-Solis, and our collaborators Laura Corbo, Susan Ledbetter, David Ledbetter, and Thomas Webster. We thank Al Edwards and Wilhelm Ansorge for unpublished sequence data, and Donna Muzny and Richard Gibbs for oligonucleotide synthesis. We also thank Michael Weil for many useful discussions. This work is funded by the Howard Hughes Medical Institute and a grant from the U.S. Department of Energy (#DE-FG05-88ER60692).

REFERENCES

1. This volume.
2. Gusella, J.F., Keys, C. Varsanyi-Breiner, A., Kao, F-T., Jones, C., Puck, T.T., and Housman, D. (1980) *Proc. Natl. Acad. Sci. USA* 77:2829-2833.

3. Burke, D.T., Carle, G.F., andd Olson, M.V. (1987) *Science* 236:806-812.
4. Britten, R.J., Baron, W.F., Stout, D.B., and Davidson, E.H. (1988) *Proc. Natl. Acad. Sci. USA* 85:4770-4774.
5. Kariya, Y., Kato, K., Hayashizaki, Y., Himeno, S., Tarui, S., and Matsubara, K. (1987) *Gene* 53:1-10.
6. Nussbaum, R.L., Airhart, S.D., and Ledbetter, D.H. (1986) *Am. J. Med. Genet.* 23:457-466.
7. Jelinek, W.R., and Schmid, C.W. (1982) *Ann. Rev. Biochem.* 51:813-844.
8. Nelson, D.L., Ledbetter, S.A., Corbo, L., Victoria, M.F., Ramirez-Solis, R., Webster, T.D., Ledbetter, D.H., and Caskey, C.T., submitted for publication.
9. Chang, S.M.W., Wager-Smith, K., Tsao, T.Y., Henkel-Tigges, J., Vaishnav, S., and Caskey, C.T. (1987) *Mol. Cell. Biol.* 7:854-863.
10. Patel, P.I., Framson, P.E., Caskey, C.T., and Chinault, A.C. (1986) *Mol. Cell. Biol.* 6:393-403.
11. Kim, S.H., Moores, J.C., David, D., Respess, J.G., Jolly, D.J., and Friedmann, T. (1986) *Nucl. Acids Res.* 7:3103-3118.
12. Triglia, T., Peterson, M.G., and Kemp, D.J. (1988) *Nucl. Acids Res.* 16:8186.

CHAPTER 12

A New Approach to Constructing Genetic Maps: PCR Analysis of DNA Sequences in Individual Gametes

Norman Arnheim

INTRODUCTION

The construction of genetic maps in higher organisms depends upon the ability to analyze the progeny of selected matings or to compute linkage relationships by means of pedigree analysis. In humans, only the latter is possible. Using restriction fragment length polymorphisms (RFLPs), there has been significant progress towards the construction of a human linkage map (for reviews see 1-3). To locate genes with known phenotypic effects relative to RFLP markers, there has been a concerted effort to establish a panel of genetic markers at about 10-cM intervals (on the average, 1 cM = 1% recombination) so that no gene will be further than 5 cM away from an RFLP marker.[4] Pedigree analysis is thought to be able to measure genetic distances to a resolution of approximately 1 cM (encompassing about 1000 kb of DNA) with statistical reliability.[5] The analysis of smaller genetic distances requires an examination of such a large number of individuals from informative families that it is impractical.

Recently, we proposed that recombination frequencies between genetic markers could be measured using a method that does not rely on genetic crosses or pedigree analysis.[6]   Our approach is based on directly determining whether individual gametes (sperm) are parental or recombinant in genotype.   This could be accomplished by determining which of two alleles are present at each of two genetic loci present in individual sperm from a doubly heterozygous man.   Dividing the number of recombinant sperm determined by this procedure by the total number of sperm examined will estimate the frequency of recombination.   The *in vitro* gene amplification procedure known as the polymerase chain reaction[7-9] is capable of amplifying DNA sequences on the order of 10^9 fold, thus raising the possibility that a single molecule of target DNA present in a single sperm could be analyzed by application of this technique.   Studying thousands of human sperm in this way would allow one to generate fine structure genetic maps at a resolution far greater than that possible using pedigree analysis.   Such an approach would provide a unique tool for the study of many problems in human genetics currently considered intractable.

PRINCIPLE OF SPERM TYPING USING PCR

From the point of view of human pedigree analysis, every individual is the result of the fusion of a single meiotic product from each parent.   Studies of the somatic DNA sequences of the parents and children combined with the appropriate statistical methods[10] allows one to deduce the recombinant or non-recombinant nature of the meiotic products that formed the zygote and to estimate what fraction of the meiotic products were recombinant.   Our alternative approach to studying genetic recombination is to determine what fraction of the meiotic products are recombinant by analyzing the DNA sequences in the meiotic products themselves.   Each man produces vast numbers of meiotic products, sperm, throughout his lifetime providing a virtually unlimited supply for direct genetic analysis.

Assume a man is heterozygous at two linked loci.   The example shown in Figure 1 indicates that the polymorphic difference at each locus is due to an AT to GC substitution.   In this example, sperm derived from this individual will be either a parental type (AT-AT or GC-GC) or a recombinant type (AT-GC or GC-AT) at the polymorphic sites of these two loci.   The frequencies of the recombinant types will depend upon the genetic distances of the two loci from each other.   The frequency of each of these four sperm types could be accurately determined by examining a large number of single sperm from this individual.   The technical problem is to determine which of the two alleles present at each locus in the premeiotic cell are in each individual sperm.   Standard molecular biological techniques are not sensitive enough to allow the analysis of the single molecule of each chromosomal DNA present in the sperm.

Using PCR to amplify the polymorphic region at each locus would be the first step in analyzing single sperm (Figure 2).   Two sets of primers, one for each locus, are required and must flank the polymorphic region.

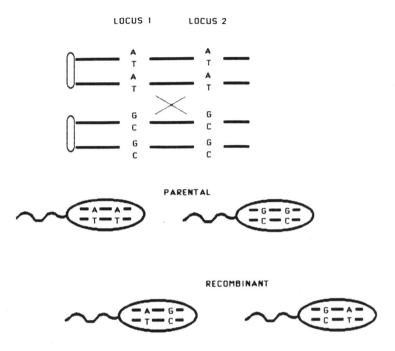

Figure 1. The meiotic products expected from a recombination event between two linked loci. A pair of homologous chromosomes consisting of four chromatids are shown at metaphase of meiosis. The nucleotide in both DNA strands are shown at the polymorphic region of each chromatid. A recombination event between the inner two chromatids would, after meiosis, yield two parental and two recombinant sperm. In each sperm, the nucleotides present in each strand at both loci are indicated.

Once the two target sequences in a sperm are amplified by PCR, the allelic composition at each locus must be determined. A procedure utilizing synthetic oligonucleotide probes has been developed to distinguish between alleles that differ by as little as a single nucleotide substitution. The first use of these allelic-specific oligomer (ASO) probes was to distinguish between the normal βA allele and the sickle cell (βS) mutation in the human β-globin gene.[11] The method requires a small oligonucleotide (typically 19 bp) to be synthesized for each allele to be tested. Each oligonucleotide matches one allele exactly but differs from the other allele usually by a single base change. The radiolabeled oligonucleotides are each separately hybridized to an aliquot of the amplified DNA sample to be typed. Appropriate hybridization conditions are used so that each oligonucleotide will form a stable duplex only if perfectly complementary sequences are present in the sample. For example, in Figure 3, the PCR products from a single sperm are analyzed with four ASO probes. One pair

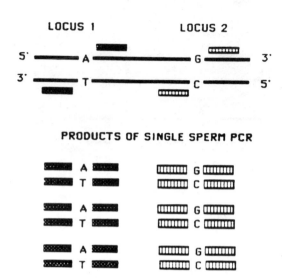

Figure 2. PCR products expected from the simultaneous amplification of the polymorphic regions of locus 1 and locus 2 in a single sperm which resulted from the recombination event shown in Figure 1.

of ASOs will distinguish between the two alleles at locus 1 and the other pair between the alleles at locus 2. ASOs allow one to be sure that each sperm examined is, in fact, haploid for the chromosome being studied since we expect only one of the two ASOs to hybridize to any one meiotic product.

One haploid human cell should contain 1.5×10^{-24} moles of any unique DNA sequence. From previous data, we know that 1 femptomole (10^{-15} moles) of a particular DNA sequence can be detected in less than a day with radioactive probes labeled at a specific activity of 5 μCi/picomole.[7] Thus, the target DNA needs to be amplified by a minimum of 10^9-fold ($10^{-15}/10^{-24}$). Even with a low average amplification efficiency of 50%, this can be accomplished with about 50 amplification cycles.

PCR ANALYSIS OF SINGLE CELLS

Single Diploid Cells

Before analyzing single sperm, Li *et al.*[6] studied the human β-globin genes in individual diploid cells. Two tissue culture cell lines were used for these experiments. One was derived from an individual homozygous for the sickle cell mutation at codon 6 (βS) and the other was homozygous for the normal βA allele. PCR primers that amplify a β-globin fragment containing codon 6 and allele specific oligonucleotide probes (ASO) capable of

PRODUCTS OF SINGLE SPERM PCR

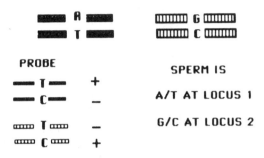

Figure 3. Analysis of the genotype of the recombinant sperm shown in Figure 2. The PCR products shown in Figure 2 can be analyzed with two pairs of ASO's, one for each locus. + and - represent positive and negative hybridization. The deduced genotype of this sperm is also shown.

distinguishing between these two alleles have already been described.[7,12] Cells homozygous for βA and cells homozygous for βS were co-cultivated in the same tissue culture flask for several days. Individual cells from this mixture were drawn into a thin plastic pipette while being observed under a phase contrast microscope. Each individual cell was delivered into a PCR tube containing a lysis solution and after incubation, PCR buffer containing dNTPs, *Taq* DNA polymerase[9] and a set of PCR primers that amplify the informative region of the globin gene were added. After 10 min at 95°C to denature the target DNA, 50 cycles of amplification were carried out according to a modification of a published procedure.[9] The details of these methods are described in Li *et al.*[6] and the legend to Figure 4. Aliquots from each sample of amplified product were hybridized separately with the βA and βS probes after fixation to nylon membranes by dot-blot hybridization.[11,12] The data are shown in Figure 4. Among the 37 cells analyzed, 84% hybridized with only one of the two allele specific probes and with varying intensities; 19 hybridized with the βA probe and 12 hybridized with the βS probe. None of the 12 control tubes that received water in place of a cell were positive indicating that DNA contamination was insignificant. No sample hybridized with both probes indicating that only a single cell was introduced into each tube and that DNA from lysed βA or βS cells present in the co-cultivation tissue culture media did not adhere to individual cells.

The amount of amplified β-globin gene product produced by PCR of a single cell was determined by comparing the intensity of the sample hybridization signal with that from known amounts of plasmid DNA carrying the globin gene that were spotted on the same filter. It was

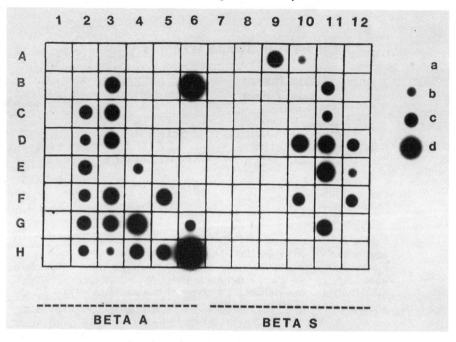

BETA A　　　　　　　**BETA S**

Figure 4. PCR analysis of the β-globin gene in individual tissue culture cells. Each of two aliquots from the PCR products of a single cell were placed in the same row of the dot-blot apparatus separated by 6 columns. Half of the filter was hybridized to the βA ASO and the other to the βS ASO. 1A-1H (7A-7H) and 2A,2B (8A,8B) were water blanks. 6H and 12H are aliquots of PCR product of purified βA DNA. One, 3, 10 and 30 femtomoles of βS gene containing plasmid were dotted as hybridization standards at positions a, b, c, and d, respectively. The remaining samples were individual tissue culture cells. Each single cell sample obtained by micromanipulation was delivered into a 0.5 ml plastic microfuge tube containing 10 μl of autoclaved distilled water. The lysis procedure is discussed in the text. After 1 hr at 37°C, samples were heated at 85°C and suspended in 100 μl of PCR reaction mix containing 1 X PCR buffer, 1 μM of each oligonucleotide PCR primer, 187.5 μM each of dATP, dCTP, dGTP, dTTP, 100 ng of *E. coli* DNA, 2 units of *Thermus aquaticus* thermostable DNA polymerase (Perkin-Elmer Cetus Instruments), and 60 μl of mineral oil to prevent evaporation. The PCR reactions were carried out on a DNA Thermal Cycler (Perkin-Elmer Cetus Instruments). Each cycle of PCR consisted of incubation at 95°C for 15 sec, 15 sec incubation at 54°C and a 1-min incubation at 72°C. After 50 cycles of PCR, dot-blot analysis of 20 μl samples of the PCR reaction were carried out using βS and βA allele specific probes.[12] *(Reprinted by permission from Nature, Vol. 335, pp. 414-417. © 1989 Macmillan Magazines Ltd.)*

estimated that starting with a diploid amount of globin DNA (3.0×10^{-24} moles) between 5 and 500 femtomoles of PCR product was produced in 50 cycles which is equivalent to an average amplification yield of 7.6×10^{10}-fold with an average efficiency per cycle of 65%. Considering the extent of amplification, elimination of all possible sources of contamination is critical to the success of all single cell experiments.

Analysis of Single Sperm

Li *et al.*[6] next analyzed the genotype of single sperm derived from an individual heterozygous at the gene which codes for the LDL receptor (LDLr) which has been localized to chromosome 19.[13] They adapted the detection of an LDLr RFLP[14] to PCR and ASO analysis with the help of unpublished DNA sequence information (David Russell, unpublished data). The PCR primers and probes have been described previously.[6]

Sperm were purified from a semen sample by centrifugation through a sucrose step gradient (see ref. 6) and stored for 8 months at -20°C. Individual sperm were drawn into a fine plastic needle under microscopic observation and delivered to a tube for lysis. Lysis was achieved by using a slight modification of a published method.[15] The sperm in 10 µl of water was adjusted to a final volume of 20 µl containing 1 X PCR reaction buffer (50 mM KCl, 10 mM Tris.HCl pH 8.3, 2.5 mM $MgCl_2$, 0.1 mg/ml gelatin), 0.05 mg/ml Proteinase K, 20 mM DTT, and 1.7 µM SDS. After 1 hr at 37°C, samples were heated at 85°C before the PCR reagents were added. The amplification protocol is detailed in the legend to Figures 4 and 5. The LDLr genotypes in 80 individual sperm were analyzed. Typical results from one such experiment are shown in Figure 5. Altogether, 55% of the sperm gave a hybridization signal. Twenty-two carried the LDLr1 allele; 21 the LDLr2 allele. Only one sample was positive with both probes. Sixteen additional control tubes which received all of the reagents but no sperm did not give any hybridization signal. The distribution of the two amplified alleles obeyed Mendel's law of independent segregation indicating that the PCR reactions were initiated with a single meiotic product and that no contaminating diploid DNA sequences were present.

Li *et al.*[6] next simultaneously amplified DNA sequences at two different loci in a single sperm which would be essential for carrying out recombination studies. Their sperm donor was heterozygous at the HLA DQα locus (DQα; see ref. 6 for details on probes and primers) on chromosome 6 as well as the LDLr gene on chromosome 19. Initial attempts to amplify both loci while both LDLr and DQα primer pairs were present throughout the entire amplification experiment were unsuccessful. Instead only the first 20 amplification cycles were carried out in the presence of both primer pairs. After this primary amplification, 1/50 of the reaction mixture was placed in a tube and diluted with a PCR solution containing the HLA primers only and another aliquot was diluted with a PCR mix containing only the LDLr primers. After an additional 45 cycles of amplification, part of each secondary reaction was hybridized to either of

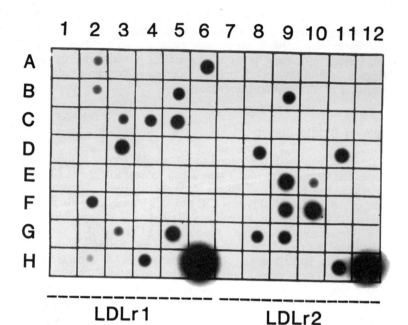

Figure 5. PCR analysis of the LDL receptor locus of individual human sperm. 1A-1H (7A-7H) are 6 water blanks. 6H and 12H are aliquots of amplified DNA from an LDLr1/LDLr2 heterozygote. The remaining samples are from individual sperm. The organization of the samples on the dot-blot is as in Figure 4. The details of sperm purification can be found in Li *et al.*[6] Single sperm were isolated in the same way as individual diploid cells using a sperm suspension at a concentration of 1×10^5 sperm/ml. The PCR and dot-blot analysis were carried out as described in Figure 1 except that the final washes of the filters hybridized with the LDLr probes were at 56°C. *(Reprinted by permission from Nature, Vol. 335, pp. 414-417. © 1989 Macmillan Magazines Ltd.)*

the two ASOs specific for the alleles at that locus.   A total of 150 individual sperm were analyzed in a series of such experiments and a summary of the data is shown in Table 1.   Hybridization signals were detected in 123 samples (82%).   The remaining 27 samples did not exhibit amplification of either locus.   In 9 of the 123 samples, two alleles from at least one of the two loci were detected and these samples were excluded from further analysis.   These samples probably contained two sperm of different genotypes rather than having resulted from meiotic non-disjunction events since recent cytological data on human sperm suggest that the average frequency of non-disjunction for a single chromosome is on the order of 0.1%[16]; far below the frequency observed.

Table 1.

Total Number of Sperm Examined	150
No signal	27
DQA1,LDLr1	21
DQA1,LDLr2	18
DQA2,LDLr1	14
DQA2,LDLr2	17
DQA1	14
DQA2	4
LDLr1	10
LDLr2	16
DQA1,LDLr1,LDLr2	2
DQA2,LDLr1,LDLr2	1
DQA1,DQA2	1
LDLr1,LDLr2	2
DQA1,DQA2,LDLr1,LDLr2	3
Controls	32
No Signal	29
LDLr1	2
LDLr2	1

Summary of data obtained by typing single sperm for polymorphisms at the LDLr and DQα loci simultaneously.[6] LDLr1 and LDLr2 refer to the alternate alleles at the LDLr locus while DQA1 and DQA2 are alleles of DQα.

Among the remaining 114 sperm, 96 could be typed at the LDLr locus with 45 having LDLr1 and 51 having LDLr2. The two alleles assorted independently at the expected 1:1 ratio (Chi square = 0.375, 0.75> P >0.5, 1 df). Eighty-eight could be typed at the DQα locus: 53 had the DQA1 allele, 35 the DQA2. The segregation of the DQα alleles at the expected 1:1 ratio is at the borderline of statistical significance (Chi square = 3.68, 0.1> P >.05, 1 df). This could represent 1) a statistical fluctuation, 2) unequal amplification of the two DQα alleles resulting from base mismatches between the PCR primers and the DQα DNA sequence of the particular donor due to the extensive population polymorphism of this locus, or 3) some unusual genetic phenomenon such as segregation distortion.

Seventy out of the 114 sperm (61%) that gave hybridization signals could be typed at both loci. The independent segregation of chromosome 6 and chromosome 19 should result in the equally frequent occurrence of the four possible gametes: DQA1,LDLr1; DQA1,LDLr2; DQA2,LDLr1; DQA2,LDLr2. Li *et al.*[6] actually observed 21, 18, 14 and 17 of each type, respectively, which is not significantly different from the expected distribution (Chi square = 1.43, 0.75> P >0.5, 3 df). These results showed that one can reliably and efficiently determine the genotype of individual sperm at two distinct genetic loci simultaneously.

Among the 114 samples that gave a hybridization signal, 18 showed one of the two DQα alleles but an LDLr product could not be detected. Twenty-six sperm showed one of the two LDLr alleles but no amplification of the HLA locus. These sperm may have been nullosomic for one of the two chromosomes studied but because of what is known about the frequency of non-disjunction (see above) failure of amplification is the most likely alternative.

The relative frequency of successful amplification of the LDLr and DQα loci in single sperm is approximately the same. Among the 141 samples, 88 amplified DQα and 96 amplified LDLr. Thus, the probability of detecting DQα and LDLr amplification products was 62% and 68%, respectively. This is only an estimate since one cannot be sure that every one of the 141 samples did, in fact, contain a sperm. However, using this estimate, one can calculate the expected frequency of samples with 1) a single allele amplified at both loci, 2) only one locus, or 3) neither locus amplified under the assumption that amplification at each locus is an independent event. Altogether, 69 sperm showed amplification of a single allele at both loci, 45 a single allele at only one locus, and 27 amplified neither locus. When this observation was compared to the expected frequencies of each class based upon independent amplification, the result was highly statistically significant (Chi square = 13.0, P >0.995, 2df). There was an excess of sperm that showed amplification of both loci or neither locus and far fewer than expected which amplified only one of the two loci. These data are consistent with the idea that after lysis a particular sperm is either a good substrate for amplification of both loci or it is bad for both as would be the case if the sperm did not undergo any lysis. Thus, amplification at both loci may not be independent events if sperm lysis is not uniform from sample to sample. Therefore, the accessibility of one of the target DNA molecules to the PCR reagents might be positively correlated with the accessibility of the other target. Improving the absolute rate of successful amplification of single sperm might result from improving lysis procedures thereby enhancing target DNA accessibility.

The Problem of "False Recombinants"

The experiments described above were designed to demonstrate the feasibility of single sperm typing and used unlinked genetic markers. In a linkage study, the frequency of recombination would theoretically be directly calculated from single sperm typing experiments by dividing the number of recombinant sperm observed by the total number of recombinant and non-recombinant sperm examined. In practice, however, one must take into consideration an artifact which can result from the simultaneous occurrence of two experimental errors. The first is that a sperm sample may contain not one sperm but two. The second is that not all of the target DNA sequences present in the sample may be amplified by PCR. Consider the situation where a male is heterozygous at two loci with alleles A and B on one chromosome and a and b on the other (the phase). If instead of one sperm, two (one of each non-recombinant type) are in the sample and only one of the two alleles at each locus is amplified by PCR (denoted by brackets in the figure below) then in the example below the sample will be typed incorrectly as an aB recombinant.

$$\text{---A--------------[B]---}$$
$$\text{--[a]--------------b---}$$

The frequency of these "false recombinants" will be a function of 1) the frequency of samples containing double sperm, and 2) the probability of amplifying an allele to a detectable level. The lowest frequency of recombination that can be measured with statistical reliability, therefore, is limited by the frequency of these "false recombinants." For the highest resolution genetic mapping studies, reducing the number of "false recombinants" by decreasing the number of samples containing two sperm and increasing the efficiency of lysis and detection of amplification, products from both loci will be required.

Another possible source of "false recombinants" is contamination of sperm samples by either total human DNA of exogenous origin or PCR products from previous amplifications. Special precautions can be taken to minimize this source of error.

Reducing "False Recombinants"

One of the most inconsistent aspects of the sperm typing protocol as it now stands and which contributes to "false recombinants" involves the manual isolation of single sperm. The actual number of samples showing evidence of two sperm and an estimate of the probability of detecting an allele by PCR can be used to calculate the total number of samples that must have contained two sperm. Our recent experiments (Li, Cui, and Arnheim; unpublished data) show that the frequency of samples which are calculated to contain two sperm varies significantly from experiment to experiment by at least a factor of 10. The use of flow cytometry[17] for sperm isolation instead of micromanipulation may significantly circumvent this problem by reducing this frequency of double sperm to less than 1% and do so in a manner that is consistent from experiment to experiment.

In addition to reducing the number of samples that contain more than one sperm, the number of false recombinants can be reduced by increasing the probability of amplifying a target sequence to a detectable level. For example, if this probability was 100%, then in every sample that carried two sperm, one of each non-recombinant type, all four alleles would be detected and the sample would be discarded as being uninformative. As discussed above, we believe that the experiments of Li *et al.*[6] showed a low probability (about 65%) due to inefficient sperm lysis. We have now completely changed the basis for our lysis procedure and, as a result, our probability of detection has risen from about 65% to often more than 90% for each of the two loci. The basis for this improvement (Li, Cui, and Arnheim; unpublished data) is the use of an alkali lysis step at high temperature in the presence of reducing agent followed by neutralization with Tris buffer and slightly modified amplification conditions. Although "false recombinants" could arise by contamination of sperm samples with exogenous DNA sequences, there is little evidence that this is a significant problem in our recent experiments. None of 63 control samples that received water instead of a sperm gave a PCR signal with any one of the four ASOs tested (Li, Cui, and Arnheim; unpublished data). We minimize contamination by storing all reagents in small aliquots that are used only once and by setting up PCR experiments in a laminar flow hood using pipets restricted for this purpose. In conclusion, by combining flow cytometry for sperm isolation and the new sperm lysis procedure, we estimate that we should be able to increase the resolution of gene mapping to less than 0.01% recombination.

APPLICATIONS OF SINGLE SPERM TYPING

Constructing Genetic Maps

Three Point Crosses. The ability to type individual sperm at the DNA level will provide a fundamentally new and alternative approach to determining the physical order of DNA polymorphisms which are so tightly linked (less than 1% recombination) that they cannot be resolved by pedigree analysis. This may be especially significant in the case of random polymorphisms tightly linked to disease-causing loci. Because this method will allow the examination of very large numbers of meiotic products, even tightly linked polymorphisms could be ordered with respect to one another by three point crosses. Such fine structure maps might prove of great value in attempts to locate the disease causing locus itself. Of course, since sperm do not exhibit disease phenotypes, an unknown disease locus cannot be directly mapped relative to DNA polymorphisms in this way.

The linear order of genetic markers on a chromosome can be inferred by carrying out three point crosses according to classical procedures used in experimental genetics.[18] Assume that a man is triply heterozygous at three closly linked loci, A, B and C on a chromosome with codominant alleles a, b and c and with the order unknown. Using the single sperm approach, the phase of these markers can be deduced since, with tightly linked polymorphisms, the most common sperm types will be non-recombinant. The order can also be determined and is based upon observing which two among the eight possible meiotic products are the least common. Thus, if ABC and abc were the most common types, these would be expected to be non-recombinant. If ABc, its reciprocal abC and AbC, and its reciprocal aBc were next most common and aBC and its reciprocal Abc were least common then these latter two products would be expected to be the result of double crossovers. These relative frequencies are what would be expected if locus A was located between B and C (BAC or CAB) and thus would establish the order.

Michael Boehnke (University of Michigan) has devised a sequential strategy for determining gene order in three point crosses using data on single sperm.[19] Boehnke's calculations, for example, show that the analysis of as few as 600 sperm could allow an accurate determination of the order of three loci if adjacent markers are as little as 0.5% recombination apart. We wish to point out that even though the absolute recombination frequencies between the same markers can differ between males and females (see ref. 2), the order of loci determined by this method will of course be the same for both sexes. So that "false recombinants" do not adversely affect the resolution of such studies (see above), the efficiency of amplification will also have to be unaffected by the presence of three primer sets during amplification. It has recently been reported that as many as seven independent segments were simultaneously amplified using large amounts (relative to a single sperm) of genomic DNA.[20] Our preliminary experiments (Li, Cui and Arnheim; unpublished data) using five sets of primer pairs simultaneously on single sperm suggest that high efficiency amplification of a particular target DNA is not affected by the simultaneous amplification of four additional loci.

Selection of Polymorphic Markers. For any polymorphism to be useful for sperm typing by PCR, DNA sequence information flanking the polymorphic site must be known. The amount of sequence information need not be more than 20-30 base pairs on either side as this would even allow some latitude in the selection of PCR primers. Unfortunately, many cDNA and genomic clones which can reveal RFLPs do not themselves contain the polymorphic site and, therefore, additional cloning and sequencing efforts would be required to adapt them to PCR analysis. In addition to the sequence information that would be required for primer synthesis, RFLPs have usually not been analyzed extensively enough so as to reveal the nature of the nucleotide substitution. In some cases, the PCR product can be directly tested for an RFLP by restriction enzyme digestion[21] although very often in single cell experiments we see multiple bands by ethidium bromide staining thereby confusing the issue. In addition, having to carry out restriction enzyme digestion and gel electrophoresis (and perhaps Southern blotting and hybridization) on a very large number of samples would be more labor intensive than carrying out dot-blot hybridization.

One particular form of RFLP, that based upon variable number of tandem repeats or VNTR, has contributed significantly to gene mapping studies using conventional methods (see 22). If 20-30 base pairs of DNA sequence information can be obtained on each side of the cluster of repeating units then, following PCR, simple gel electrophoresis followed by ethidium bromide staining might be used to distinguish between alleles. Using genomic DNA, analysis of VNTRs amplified by PCR has recently been reported.[23,24] If it were possible to eliminate the presence of multiple bands currently seen in single cell amplification experiments, these VNTR polymorphisms could be used in sperm typing studies.

Whether it is more efficient to convert known RFLP's to markers that can be typed by PCR or whether it would be advantageous to work out techniques for rapidly finding new polymorphisms more easily adaptable to PCR analysis is unclear at this point. There are, however, a number of newer strategies that can be coupled with PCR for finding polymorphisms. Leonard Lerman's denaturing gradient gel electrophoresis method (DGGE)[25] which has been shown to successfully detect single base mutations[26,27] can be applied to PCR products using the GC clamp[28] or a modification of DGGE which involves digestion with multiple restriction enzymes (Mark Grey, personal communication). Nucleotide substitutions present in PCR products may also be analyzed using RNAse A mismatch analysis.[29-31] Once a polymorphism is detected using these methods, its position can be determined with relatively little difficulty and sequence information at and around the site acquired so as to provide data for the synthesis of ASO probes. Using these new approaches, new polymorphisms within clones that have already been shown to detect RFLPs by Southern blotting and which have already been mapped might be discovered. This might significantly speed up attempts to convert RFLP's to PCR analysis and allow us to order already existing RFLP's which are closly linked to disease causing loci.

Impact on Conventional Mapping. Being able to determine the genotype of individual sperm could have a significant impact on conventional gene mapping using pedigree analysis. The maximum amount of information in family studies is obtained when the phase of the genetic markers in the parents is known[10] but frequently this information is not available. Using single sperm typing, the phase in male parents could be easily determined by observing which class of sperm are the most abundant. When the markers are reasonably close together, this will identify the non-recombinant chromosomes and thus the phase.

Studying the Relationship Between Recombination and Physical Distance

Another significant advantage of being able to type a large number of meiotic products is that it will allow accurate determination of the recombination frequency between genetic markers which are physically very close together. In conjunction with gel electrophoresis procedures for large DNA fragments and chromosome walking data, it would be possible to measure the frequency of recombination between genetic markers whose physical distance apart is known precisely. This would allow a comparison of the relationship between recombination frequency and physical distance for specific chromosomal regions with the conventionally accepted average value of 1% recombination per million base pairs.[32]

The ability to measure recombination over short physical distances will be especially useful in the study of recombination hot spots. Recombination hot spots in humans have been proposed to exist in the β hemoglobin complex, the insulin locus and the immunoglobulin heavy chain region (see refs. 33-35) based upon population linkage disequilibria data. The pseudoautosomal region of the X and Y chromosomes appears to undergo recombination at a rate 20 times the average rate of 1% per 1000 kb.[36,37] Defining specific hot spot regions within a longer DNA segment by pedigree analysis would be a prodigious task. It is clear that any new techniques that can assess recombination potential across small DNA segments will prove valuable in attempts to understand the molecular mechanisms behind recombination enhancement, and may contribute to our understanding of the recombination process itself.

Other Applications

Because it should be possible to obtain statistically significant data on recombination frequencies from a single individual, it should also be possible to determine whether different males have the same or different rates of recombination for the same chromosomal interval and whether the rate for a specific interval changes with age (see ref. 38). This could provide information which would be valuable for genetic counseling.

The typing of single sperm in organisms that cannot be bred extensively or which have exceptionally long generation times may be the only practical way of making genetic maps for these species. This could be especially significant in studies on primates such as chimps and gorillas where physical mapping data in conjunction with single sperm recombination studies might provide us with information on the evolution of recombination potential and the role that genetic variability generated by recombination may have played in human evolution.

It is also of some significance that the ability to study DNA sequences in single cells other than sperm will make it possible to study cell to cell variation in developmental processes which involve DNA rearrangements or other genetic alterations. Prenatal diagnosis carried out on a single cell derived from a preimplantation embryo resulting from *in vitro* fertilization is also possible.

In conclusion, the ability to analyze DNA sequences in single sperm or diploid cells can provide unique approaches to problems not only in human genetics but in other areas of biology as well.

ACKNOWLEDGMENTS

The author greatly appreciates helpful discussions with Honghua Li and Mike Boehnke.

REFERENCES

1. Donis-Keller, H., Green, P., Helms, C., Cartinhour, S., Weiffenbach, B., Stephens, K., Keith, T.P., Bowden, D.W., Smith, D.R., Lander, E.S., Botstein, D., Akots, G., Rediker, K.S., Gravius, T., Brown, V.A., Rising, M.B., Parker, C., Powers, J.A., Watt, D.E., Kauffman, E.R., Bricker, A., Phipps, P., Muller-Kahle, H., Fulton, T.R., Ng, S., Schumm, J.W., Braman, J.C., Knowlton, R.G., Barker, D.F., Crooks, S.M., Lincoln, S.E., Daly, M.J., and Abrahamson, J. (1987) *Cell* 51:319-337.
2. White, R. and Lalouel, J.-M. (1987) *Advances in Human Genetics* 16:121-228.
3. White, R., Lalouel, J.-M., O'Connell, P., Nakamura, Y., Leppert, M., and Lathrop, M. (1987) *Cytogenet. Cell Genet.* 46:715.
4. Botstein, D., White, R.L., Skolnick, M., and Davis, R.W. (1980) *Am. J. Hum. Genet.* 32:314-331.
5. Aston, C.E., Sherman, S.L., Morton, N.E., Speiser, P.W., and New, M.I. (1988) *Am. J. Hum. Genet.* 43:304-310.
6. Li, H., Gyllensten, U.B., Cui, X., Saiki, R.K., Erlich, H.A., and Arnheim, N. (1988) *Nature* 335:414-417.
7. Saiki, R., Scharf, S., Faloona, F., Mullis, K.B., Horn, G.T., Erlich, H.A., and Arnheim, N. (1985) *Science* 230:1350-1354.
8. Mullis, K.B. and Faloona, F.A. (1987) *Meth. in Enzymology* 155:335-350.
9. Saiki, R.K., Gelfand, D.H., Stoffel, S., Scharf, S.J., Higuchi, R., Horn, G.T., Mullis, K.B., and Erlich, H.A. (1988) *Science* 239:487-491.
10. Ott, J. (1985) *Analysis of Human Genetic Linkage.* The Johns Hopkins University Press, Baltimore, pp.172-174.
11. Conner, B.J., Reves, A.A., Morin, C., Itakura, K., Teplitz, R.L., and Wallace, B. (1983) *Proc. Natl. Acad. Sci. USA* 80:278-282.
12. Saiki, R.K., Bugawan, T.L., Horn, G.T., Mullis, K.B., and Erlich, H.A. (1986) *Nature* 324:163-166.
13. Lindgren, V., Luskey, K.L., Russell, D.W., and Francke, U. (1985) *Proc. Natl. Acad. Sci. USA* 82:8567-8571.
14. Hobbs, H.H., Esser, V., and Russell, D.W. (1987) *Nucl. Acids Res.* 15:379.
15. Gill, P., Jeffreys, A.J., and Werrett, D.J. (1985) *Nature* 318:577-579.
16. Brandriff, B., Gordon, L., Ashworth, L.K., Watchmaker, G., and Carrano, A.V. (1986) in *Genetic Toxicology of Environmental Chemicals, Part B: Genetic Effects and Applied Mutagenesis.* Alan R. Liss, Inc., pp. 469-476.
17. Van Dilla, M.A. *et al.* (1977) *J. of Histochemistry and Cytochemistry* 25:763-773.
18. Sturtevant, A.H and Beadle, G.W. (1939) *An Introduction to Genetics.* Dover Publications, Inc., N.Y. Chapter 6.
19. Boehnke, M., Arnheim, N., Li, H., and Collins, F. (1989) *Am. J. Hum. Genet.* In press.
20. Chamberlain, J.S., Gibbs, R.A., Ranier, J.E., Nguyen, P.N., Farwell, N.J., and Caskey, C.T. (1988) *NAR* 16:11141-11155.
21. Kogan, S.C., Doherty, M., and Gitschier, J. (1987) *N. Engl. J. Med.* 317:985-990.
22. Nakamura, Y., Leppert, M., O'Connell, P., Wolff, R., Holm, T., Culver, M., Martin, C., Fujimoto, E., Hoff, M., Kumlin, E., and White, R. (1987) *Science* 235:1616-1622.
23. Jeffreys, A.J., Wilson, V., Neumann, R., and Keyte, J. (1988) *NAR* 16:10953-10971.
24. Weber, J., and May, P.E. (1988) *Am. J. Hum. Genet.* 44:388-396.

25. Fischer, S.G., and Lerman, L.S. (1983) *Proc. Natl. Acad. Sci. USA* 80:1579-1583.
26. Cariello, N.F., Scott, J.K., Kat, A.G., Thilly, W.G., and Keohavong, P. (1988) *Am. J. Hum. Genet.* 42:726-734.
27. Myers, R., and Maniatis, T. (1986) *Cold Spring Harbor Symp. Quant. Biol.* 51:275-283.
28. Sheffield, V.C., Cox, D.R., Lerman, L.S., and Myers, R.M. (1989) *Proc. Natl. Acad. Sci. USA* 86:232-236.
29. Myers, R.M., Larin, Z., and Maniatis, T. (1985) *Science* 230:1242-1246.
30. Winter, E., Yamamoto, F., Almoguera, C., and Perucho, M. (1985) *Proc. Natl. Acad. Sci. USA* 82:7575-7579.
31. Almoguera, C., Shibata, D., Forrester, K., Martin, J., Arnheim, N., and Perucho, M. (1988) *Cell* 53:549-554.
32. Renwick, J.H. (1969) *Br. Med. Bull.* 25:65-73.
33. Chakravarti, A., Buetow, K.H., Antonarakis, S.E., Waber, P.G., Boehm, C.D., and Kazazian, H.H. (1984) *Am. J. Hum. Genet.* 36:1239-1258.
34. Chakravarti, A., Elbein, S.C., and Permutt, M.A. (1986) *Proc. Natl. Acad. Sci. USA* 83:1045-1049.
35. Migone, N., de Lange, G., Piazza, A., and Cavalli-Sforza, L.L. (1985) *Am. J. Hum. Genet.* 37:1146-1163.
36. Page, D.C., Bieker, K., Brown, L.G., Hinton, S., Leppert, M., Lalouel, J.-M., Lathrop, M., Nystrom-Lahti, M., De La Chapelle, A., and White, R. (1987) *Genomics* 1:243-256.
37. Petit, C., Levilliers, J., and Weissenbach, J. (1988) *EMBO J.* 7:2369-2376.
38. Weitkamp, L.R., van Rood, J.J., Thorsby, E., Bias, W., Fotino, M., Lawler, S.D., Dausset, J., Mayr, W.R., Bodmer, J., Ward, F.E., Seignalet, J., Payne, R., Kissmeyer-Nielsen, F., Gatti, R.A., Sachs, J.A., and Lamm, L.U. (1973) *Hum. Hered.* 23:197-205.

CHAPTER 13

Evolutionary Analysis via PCR

Thomas D. Kocher and Thomas J. White

Evolutionary genetics has two parallel thrusts: the reconstruction of phylogenies and the analysis of populations. Since 1962, when Zuckerkandl and Pauling[1] proposed that comparison of the sequences of proteins and genes could be used as a molecular clock to date the divergence of extant species, a variety of biochemical methods has been used for phylogenetic studies. Electrophoretic analysis of isozymes, immunological comparisons, and protein sequencing were extensively used in the first two decades and, more recently, DNA-DNA hybridization and ribosomal RNA sequencing have made a major contribution to systematics.[2] Most of these techniques are limited because they estimate sequence difference instead of measuring it directly.

Population biologists, who have concentrated on isozyme analyses and restriction mapping of nuclear and organelle DNAs,[3,4] need methods with a greater resolving power. Until now, it has been impractical to obtain sequence data on more than just a few model organisms. The effort required to screen clone libraries and map and sequence clones has been far too great to examine more than a few individuals of a given species. The polymerase chain reaction and direct sequencing overcome the limitations of traditional DNA methods for evolutionary studies.[5]

UNIVERSAL PRIMERS

At first glance, it would appear that the lack of sequence knowledge for most species would limit the application of the PCR, since some knowledge of the sequence is required to design primers for PCR. Fortunately, there is a rapid means of obtaining primer sequences for new species. By choosing sequences that are conserved among widely divergent species, "universal" primers can be designed that will amplify a particular nuclear or organelle gene fragment from nearly all members of a major taxonomic group, e.g., plants or fungi. This extends the phylogenetic range of comparative sequence analyses to the taxonomic level of class or phylum. But the method also has important uses for identifying specimens and scoring types in population work. For these purposes, mitochondrial DNA (mtDNA) sequences are often appropriate. Below, we describe "universal" primers for both nuclear and mitochondrial sequences. Mitochondrial genes are a rigorous test of the method; since mtDNA evolves so quickly, it should be difficult to design primers for this molecule.

Primers for Nuclear Genes

The concept of "universal" primers has been used for a number of years in the direct sequencing of abundant RNAs.[6] Those same primers are easily paired to amplify ribosomal DNA sequences via PCR. Amplification of the rDNA sequences offers several advantages over direct rRNA sequencing. First, DNA can be used as the starting material, and this is usually easier to prepare from tissues than is RNA. Second, much smaller samples can be used.[7] Finally, DNA sequencing methods can be used — including a variety of sequencing enzymes and techniques — to obtain data from both strands and to sequence through the complex secondary structures found in ribosomal RNAs.

As an example, we have designed primers that amplify a region of approximately 515 bp in size of 18S rDNA from many fungi, protozoans, algae, plants and animals (the size of the amplified region plus the primers is approximately 555 bp). The primers NS1 and NS2 were based on conserved nucleotide sequences among the 18S rDNA from *Saccharomyces cerevisiae*, *Dictyostelium discoideum*, and *Stylonichia pustulata* and have been described elsewhere.[8]

> NS1   5'-GTAGTCATATGCTTGTCTC-3'
>
> NS2   5'-GGCTGCTGGCACCAGACTTGC-3'

NS1 and NS2 do not amplify bacterial or mitochondrial rRNA genes using the conditions described below. Sequence differences in the region amplified by NS1 and NS2 (and for some organisms, length differences) may be useful for initial estimates of molecular diversity, i.e., determining the number of different species in a natural population of various organisms. These primers have also found use in detecting the presence of fungal DNA contamination in ancient DNA isolated from plant and animal tissues. Medlin and coworkers[9] describe other rDNA primers that amplify the entire 18S gene from lower eukaryotes; these primers will be more useful than NS1 and NS2 for phylogenetic studies of these organisms.

Primers for Mitochondrial DNA

Studies of mitochondrial gene sequences are appropriate for many problems in evolutionary and population biology.[3,5] Because of its clonal inheritance in vertebrates, it is ideal for reconstructing maternal phylogenies. Its high rate of point-mutational evolution makes it ideal for high-resolution population studies within species and the rapid fixation of mutations within species makes this molecule ideal for species identification, especially in small organisms.

Because vertebrate mtDNA evolves so quickly (roughly 10 times as fast as nuclear genes), it should be difficult to find conserved sequences to use as priming sites for the PCR. The location of two primers that amplify a 307-bp region (the amplified fragment is 376 bp) of the cytochrome *b* gene from most vertebrates tested (mammals, birds, amphibians, reptiles, and fishes) is shown in Figure 1. Primers L14841 and H15149 were based on conserved regions of published nucleotide sequences. The letters L and H refer to the light and heavy strands of mtDNA and the number refers to the position of the 3' base of the primer in the complete mtDNA sequence reported for a human.

L14841 5' -AAAAAGCTTCCATCCAACATCTCAGCATGATGAAA-3'

H15149 5' -AAACTGCAGCCCCTCAGAATGATATTTGTCCTCA-3'

We have also had success using truncated versions of these primers to amplify mammal, bird, and amphibian DNA (C. Orrego, pers. communication). Primers MVZ3 and MVZ4 have the sequence underlined above and amplify a 311-bp region (the amplified fragment is 366 bp).

The sequences from the mitochondrial cytochrome *b* gene contain phylogenetic information of high resolving power and great taxonomic range. Alignment of sequences is facilitated because of the overall conservation of the protein structure and function. Close relationships can be assessed through changes due to transitions at "silent" sites in codons; more distant relationships can be examined by analyzing transversion differences or amino acid replacements.[5]

METHODS

DNA Preparation

DNA was extracted from tissues by digestion in 100 mM Tris pH 8.0, 10 mM EDTA, 100 mM NaCl, 0.1% SDS, 50 mM DTT, 0.5 µg/ml Proteinase K

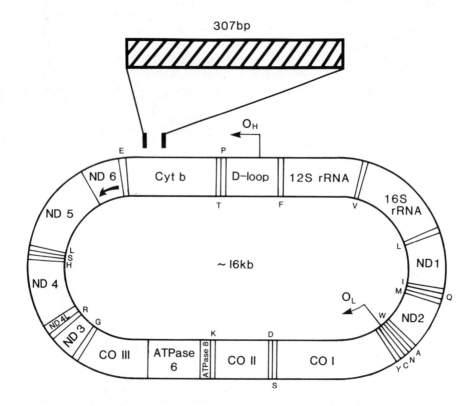

Figure 1. Position of conserved priming sites for cytochrome
b in the consensus structure of the vertebrate mitochondrial
genome. The molecule encodes 13 proteins, 2 ribosomal
RNAs, and 22 tRNAs. The regulatory region, also known as
the D-loop, is found between the genes for cytochrome *b*
and the 12s rRNA. Kocher *et al.*[5] discuss several other
conserved primer regions for mtDNA.

for several hours at 37°C. The DNA was purified by extracting twice with
phenol, once with phenol/chloroform (1:1), and once with chloroform. The
sample was then concentrated by centrifugal dialysis or ethanol precipitation.

Amplification and DNA Sequencing

Amplification was performed in 100 µl of a solution containing 67 mM
Tris pH 8.8, 6.7 mM MgSO$_4$, 16.6 mM ammonium sulfate, 10 mM β-
mercaptoethanol, 1 mM of each dNTP, 1 µM of each primer, 10-1000 ng

of genomic DNA and 2.5 units of *Taq* polymerase (Perkin Elmer/Cetus). Each cycle of the polymerase chain reaction consisted of denaturation for 1 min at 93°C, hybridization for 1 min at 50°C, and extension for 2-5 min at 72°C. This cycle was repeated 25-40 times depending on the initial concentration of template DNA in the sample. Electrophoresis of 5 μl of the amplified mixture was done in a 2% agarose gel (NuSieve, FMC Corp.) in 40 mM Tris-acetate (pH 8.0) and the DNA stained with ethidium bromide.

The gel fragment containing the amplified product was cut from the gel and melted in 1 ml of distilled water, and 1 μl of this mixture was used as the template in a second chain reaction to generate single stranded DNA for sequencing.[10] In this second reaction, the concentration of one or the other primer was reduced 100-fold. After 40 cycles of amplification, free nucleotides and salts were removed by 2-4 cycles of centrifugal dialysis (Centricon 30, Amicon). The DNA was sequenced with a commercial kit (Sequenase, U.S. Biochemical) using the primer that had been limiting in the second chain reaction.

Technical Notes

It is possible to improve the specificity and yield of the amplified product by raising the temperature of the annealing step for the DNA from some organisms. This increased specificity permits single-stranded DNA template to be produced directly from the original template in a single PCR amplification of 35 cycles using a primer ratio of 50:1. Modified reaction mixtures use reduced concentrations of each dNTP (32 μM) and cycling parameters are: an initial denaturation at 93°C for 3 min followed by 35 cycles consisting of denaturation at 93°C for 25 sec, annealing at 55°C for 25 sec and extension at 72°C for 2 min. Using longer extension times at 72°C during the linear phase may increase the yield of single-stranded template for sequencing. These cycling parameters were used with primers NS1 and NS2 for fungal DNA and eliminated the additional bands observed in lane 1 in Figure 2.

For studies where the same gene is being amplified and sequenced from many different individuals or organisms, it is imperative to scrupulously avoid cross-contamination of DNA samples during DNA isolation and manipulation of amplified products. Only positive displacement pipets which have disposable tips and pistons should be used in setting up PCR reactions. Pipets that have been used with amplified DNA should never subsequently be used for DNA isolation from tissues or for DNA dilutions prior to another round of amplification. Controls that contain no DNA and utilize all other reagents, diluents, etc., should be included in every experiment to check for contamination. In extreme circumstances, it may be necessary to switch to primers which amplify a different segment of the target gene and to obtain new pipets that are dedicated to DNA isolation or setting up PCR reactions.

RESULTS

Figure 2 shows the results of amplifying an rDNA fragment using 5 ng of total DNA from various organisms with primers NS1 and NS2. A major amplified DNA product of about 560 bp is observed from all the organisms tested and small variations in length are detected. Because these primers have such broad range, they may be useful for surveys of symbiotic organisms. For example, once sequence data have been obtained for a group of mycorrhizal fungi and host plants, species-specific DNA probes could be used to survey and identify individual symbiont partners in a natural population.

The primers for mtDNA amplify the cytochrome *b* region from a wide range of species. Figure 3 shows an alignment of mtDNA sequences from the cytochrome *b* obtained from five vertebrates. The sequences illustrate both the broad utility of the primers and the colinearity of the alignment. The primers work on human, mouse and cow mtDNA despite considerable sequence mismatch (Figure 4).

Figure 2. Amplification of a region of the nuclear small subunit rRNA gene using primers NS1 and NS2. Conditions for amplification, electrophoresis, and detection by visualization of bands stained with ethidium bromide were as described in Methods except that an annealing temperature of 45°C was used. Lane 1: *Laccaria bicolor*, a mycorrhizal basidiomycete; lane 2: *Alder glutinosa*; lane 3: *Drosophila melanogaster*, lane 4: *Anelosimus eximius*, a spider; lane 5: plasmid DNA from a clone of the rDNA repeat unit from *Tyromyces unicolor*,[19] a basidiomycete; lane 6: negative control, no DNA; lane 7: blank; lane 8: molecular weight standards, PhiX174RF *Hae*III digest.

```
            T   G   L   L   F   L   A   M   H   Y   S   P   D   A   S   T   A   F   S   S   I   A   H   I   T   R
Human      ACA GGA CTA CTA TTC CTA GCC ATG CAC TAT TCA CCA GAC GCC TCA ACC GCC TTT TCA TCA ATC GCC CAC ATC ACT CGA    25
Sheep      ACA GGC CTA CTA TTT TTA GCA ATA CAC TAT ACA CCT GAC ACA GCA GCA GCA TTT TTT TCC GTA ACC CAC ATT TGC CGA    75
Chicken    ACC GGC CTA CTA TTC TAC ACA ATG CAC CTC TCC GCA ACA GCA AAC GCC GCC TTC TTC ATC GTA GCC CAC ACT TGC CGG    75
Salamander ACA GGG TTA TTA TTT ATT ACA CAT CAT CTA ATT ATT GAC ATA ATT ACA GCT ATT TTT ATT GGA GCC CAC ATC TGC CGT    75
Fish       ACA GGC CTT CTA TTC ATT GCC ATA CAC TAC AAT TCC ATC GCC GCC ACC GCC TTT TCC TCC ATT GCC CAC ATC TGT CGT    75

            D   V   N   Y   G   W   I   R   Y   L   P   H   A   N   G   A   S   M   F   F   I   C   L   F   L   L
Human      GAC GTA AAT TAT GGC TGA ATC CGC TAC CTT CAC GCC AAT TCA GGG GCC TCA ATA TTT TTT ATC TGC CTC TTC CTA CTA    50
Sheep      GAC GTA AAC TAC GGC TGA ATC CGA TAT ATA CAC GCA AAC TCA GGG GCA TTT TTT TTC TTT ATC TGC TTA CTA TTT ATG   150
Chicken    AAC CAA TAC TAC GGC TGA CTC CGG AAT ATT CTC GCC AAC TCA GGG GCC TCA TTC TTC TTC ATC TGT ATC TTC ATC CTT   150
Salamander GAT GTA AAT TAT GGT TGA CTT CGA ATT ATT ATT CAC GCA GCT TCA GCC TCA TTT TTT TTT ATT TGT ATT TTC TTC CTT   150
Fish       GAC GTC AAC TAC GGT TGA CTC CGA AAT ATG CAC GCC AAC TCC TCA GCA TCC TCC TTC TTC ATT TGC ATT TAC TAC CTC   150

            H   I   G   R   G   L   Y   Y   G   S   F   L   Y   S   E   T   W   N   I   G   I   A   H   L   L   L
Human      CAC ATC GGG CGA GGC CTA TAT TAC GGG TCA TTT CTC TAC TCA GAA ACC TGA AAC ATC GGC ATT ATC CTC CTG CTC CTT    75
Sheep      CAT GTA GGA CGA GGC CTA TAC TAC GGA TCA TAT ACC TTC ATG GAA ACA TGA AAC ATT GGA GTA ATC CTC CTA CTC TTC   225
Chicken    CAC ATC GGA CGA GGC CTA TAC TAC GGC TCC TAC CTC TAC ATG AAG ACA TGA AAC ACA GGA GTA ATC CTC CTC CTC CTC   225
Salamander CAT ATT GGA GGA CTA TAT TTA TAT GGA TCA TTA ATG TTA TAT GAA ACA TGA AAT ATT GGA GTA ATT TTA TTA TTT CTT   225
Fish       CAC ATT GGG CGA GGG TTA TAC TCC GGG TCC TAT ATG TAT AAC GAA ACC TGA AAT ATT GGA GTT ATC CTC CTC CTT CTC   225
```

Figure 3. Partial cytochrome *b* sequences of five vertebrates. Primers L14841 and H15149 were used as described in Methods. The amino acid sequence is the mitochondrial translation of the human mtDNA sequence.[18] The other sequences are from sheep (*Ovis aries*), chicken (*Gallus gallus*), Salamander (*Ambystoma tigrinum*), and cichlid fish (*Julidochromis regani*).

143

AMPLIFICATION WITH
MISMATCHED PRIMER

PRIMER

Fish 5'—C C A T C C A A C A T C T C A G C A T G A T G A A A —3'

TEMPLATE

Human   • • • • • • • • • • • • • C • • • • • • • • • • •

Cow   • • • • • A • • • • • T • • • T • • • • • • • • • •

Mouse   • • • • • • • • • • • T • • • T • • • • • • • • • •

Figure 4.   Amplification with mismatched primers.   The
sequence of part of the cytochrome *b* primer L14841 is
shown.   Despite mismatches in the middle of the primer, it
can be used to amplify a variety of mammalian templates.   It
is most important that the 3' end of the primer be well-
matched to the template.

DISCUSSION

The ability to rapidly amplify sequences from virtually any species from
just a few molecules of DNA has many applications in the study of
evolution and ecology.   Here we discuss some promising uses of the
technique, and try to stimulate the invention of other applications in these
fields.

Molecular Systematics

While the advantages of nucleotide sequence data for systematic studies
have long been recognized, the accumulation of comparative sequence data
has until now been tedious.   The rapid sequencing which PCR/universal
primer technology provides will facilitate the extension of molecular
systematics to many more groups.   The homologous data that are provided
by sequences provide a common phylogenetic framework in varied
taxonomic groups.   Sequence data also provide a degree of resolution that
has not been possible with previous methods.   The PCR amplification and
direct sequencing of mitochondrial DNA has already been useful in
obtaining sequences that confirm the idea of an African root for the tree
relating human mtDNAs.[11]   In another study,[12] a phylogenetic analysis of
primate class II histocompatibility genes demonstrated that the origin of
many of the alleles of the HLA-DQα loci predate the speciation event that
led to humans.

Because amplification can be accomplished from only a few molecules of the target sequence, many samples previously refractory to molecular analysis can be utilized. The ability to work from museum[13] and herbarium specimens[14] should promote the use of molecular data in a variety of systematic studies. Also exciting is the application of these techniques to ancient specimens.[13,15] DNA from a 7,000-year old brain preserved in a Florida peat bog was found to contain a novel type of mtDNA not yet observed among living Native Americans.[16]

Population Biology

The rapid amplification of sequences allows the serious consideration of population studies at the DNA sequence level. Single-stranded amplifications allow rapid sequencing of DNA from tens or hundreds of individuals without the tedious cloning steps required previously. Once a representative set of sequence data is available, more convenient and simple analytical methods, such as allele specific oligonucleotide probes (see Chapter 16), can be used to obtain allele frequency data.

The ability to work from museum specimens will facilitate study of gene frequencies over time. Thomas and coworkers[17] have studied a series of rodent populations through 70 years by amplifying mtDNA sequences from the skins of preserved specimens. Variation in gene frequency over such long periods has not been accessible to modern molecular techniques until now.

Ecology and Marine Biology

Traditional molecular techniques require relatively large tissue samples for analysis. In particular, many invertebrates are too small for the manipulations normally required for molecular investigations. For this reason, direct analysis of individual small organisms, organisms that are symbiotic, or those that are not easily grown in pure culture has been difficult. The polymerase chain reaction can be used to extend the range of organisms that is accessible to molecular investigations. It should now be possible to study directly the genetic structure of natural populations of single-celled organisms such as protists and algae. The method should find wide application in analyses of dispersal and recruitment, especially in marine environments. Finally, the sensitivity of the polymerase chain reaction should allow detection and identification of small numbers of single-celled organisms, symbionts and parasites, even from complex mixtures of DNA isolated from their hosts.

Non-invasive Sampling

The collection of samples for genetic analysis has frequently required sampling of blood or tissues. This has limited the use of genetic analyses in behavioral and ecological studies where disturbance of the subjects must be minimized. The ability to amplify sequences from forensic samples such as single hairs[7] offers new opportunities for behavioral scientists and conservation biologists. Non-invasive sampling should also facilitate collection of samples for studies of endangered species.

Synergism of Molecular and Organismal Studies

By opening up a new range of species to molecular analysis, we hope a productive interaction between molecular and population biologists can arise. The comparative sequences gathered for phylogenetic reconstruction, for example, may shed light on protein structure and function. Molecular study of organisms adapted to unique environments and not usually studied in the laboratory may reveal unusual molecular adaptations. Conversely, knowledge of the molecular structure of genetic variants may contribute to our understanding of how organisms are evolving.

ACKNOWLEDGMENTS

We thank Mary Ann Brow and Randy Saiki for many helpful suggestions on DNA sequencing and PCR, Monique Gardes for providing DNA from mycorrhizal fungi, Dave Irwin for sheep DNA, Leticia Aviles for spider DNA, and Allan Wilson, Norm Arnheim, Cristian Orrego, Tom Bruns, Ellen Prager and John Taylor for helpful advice and comments on the manuscript.

REFERENCES

1. Zuckerkandl, E. and Pauling, L. (1962) in *Horizons in Biochemistry*, M. Kasha, B. Pullman, eds. Academic Press, New York, pp. 189-225.
2. Wilson, A.C., Carlson, S.S., and White, T.J. (1977) *Ann. Rev. Biochem.* 46:573-639.
3. Wilson, A.C., Zimmer, E.A., Prager, E.M., and Kocher, T.D. (1989) in *The Hierarchy of Life*, Fernholm, B., Bremer, K., Jörnvall, H., eds. Elsevier, Amsterdam, p. 407-419.
4. Palmer, J.D. (1985) in *Molecular Evolutionary Genetics*, MacIntyre, R.J., ed. Plenum Press, New York, pp. 181-240.
5. Kocher, T.D., Thomas, W.K., Meyer, A., Edwards, S.V., Pääbo, S., Villablanca, F.X., and Wilson, A.C. (1989) *Proc. Natl. Acad. Sci. USA*, in press.
6. Lane, D.J., Pace, B., Olsen, G.J., Stahl, D.A., Sogin, M.L., and Pace, N.R. (1985) *Proc. Natl. Acad. Sci. USA* 82:6955-6959.
7. Higuchi, R., von Beroldingen, C.H., Sensabaugh, G.F., and Erlich, H.A. (1988) *Nature* 332:543-546.
8. White, T.J., Bruns, T., Lee, S., and Taylor, J. (1989) *PCR Protocols: A Guide to Methods and Applications*. Academic Press, NY.
9. Medlin, L., Elwood, H.J., Stickel, S., and Sogin, M.L. (1988) *Gene* 71:491-499.

10. Gyllensten, U.B., and Erlich, H.A. (1988) *Proc. Natl. Acad. Sci. USA* 85:7652-7656.
11. Kocher, T.D, and Wilson, A.C., in preparation.
12. Gyllensten, U.B., and Erlich, H.A. (1989) *Proc. Natl. Acad. Sci. USA* 86:, in press.
13. Pääbo, S. (1989) *Proc. Natl. Acad. Sci. USA* 86:1939-1943.
14. Bruns, T.D., Fogel, R., and Taylor, J.W. (1989) *Mycologia*, submitted.
15. Pääbo, S., Higuchi, R.G., and Wilson, A.C. (1989) *J. Biol. Chem.*, 264:, in press.
16. Pääbo, S., Gifford, J.A., and Wilson, A.C. (1988) *Nucl. Acids Res.* 16:9775-9787.
17. Thomas, K., Pääbo, S., Villablanca, F.X., Wilson, A.C., in preparation.
18. Anderson, S., Bankier, A.T., Barrell, B.G., de Bruijn, M.H.L., Coulson, A.R., Drouin, J., Eperon, I.C., Nierlich, D.P., Roe, B.A., Sanger, F., Schreier, P.H., Smith, A.J.H., Staden, R., and Young, I.G. (1981) *Nature* 290:457-465.
19. Kwok, S., White, T.J., and Taylor, J.T. (1986) *Exptl. Mycology* 10:196-2

PART THREE

MEDICAL APPLICATIONS

The detection of infectious disease pathogens and the identification of genetic variation associated with disease has been revolutionized by the use of specific nucleic acid hybridization probes. PCR has helped realize the potential of clinical DNA based diagnosis by producing enough of the target sequence so that simple, rapid, and robust methods for identifying it could be employed. The first application of PCR was the prenatal diagnosis of sickle-cell anemia so it is appropriate that this section begins with a discussion of monogenic genetic diseases (Chapter 14) with particular emphasis on the hereditary hemoglobinopathies. The identification of *new* mutations in X-linked genes poses a unique challenge and the use of PCR in the diagnosis of these X-linked diseases (e.g., Duchenne's Muscular Dystrophy) is discussed in Chapter 15. The "conversion" of restriction fragment length polymorphism linkage markers for cystic fibrosis to PCR markers (detected either by restriction enzyme digestion or oligonucleotide probe hybridization to the amplified fragment)[1] for pedigree analysis and carrier identification has demonstrated that PCR amplification can contribute to genetic diagnosis even if the disease locus and mutation(s) have not yet been identified.

In general, clinical diagnostics requires a simple and rapid test. In this regard, the introduction of nonradioactive oligonucleotide probes and a novel format (the "reverse dot-blot"[2]) has significantly simplified the analysis of genetic variation. The application of this approach to the genetic typing of HLA polymorphism for tissue transplantation and disease susceptibility is presented in Chapter 16. The use of this HLA typing system as well as of other PCR genetic markers in forensic analysis is the subject of Chapter 17.

The analysis of mutations in genes that effect cellular metabolism has contributed significantly to our understanding of the regulation of cell growth and differentiation. The detection of activating mutations in the *ras* proto-oncogenes and their clinical significance are discussed in Chapter 18. Specific chromosomal rearrangements are associated with several leukemias and are, therefore, useful as diagnostic markers; they can be detected by conventional cytogenetics as well as PCR. Most B-cell follicular lymphomas contain a t(14;18) chromosomal translocation which can be detected by PCR in a majority of cases because the breakpoints in most patients are clustered in a very small section of the chromosome.[3] In the case of Philadelphia chromosome positive chronic myeloid leukemia (CML), acute lymphocytic leukemia (ALL) and acute myeloid leukemia (AML), the chromosomal breakpoints are too far apart for easy PCR amplification and analysis. Fortunately, the "fusion" mRNA transcript (BCR-ABL) is unique to leukemic cells, and its detection by modified PCR reactions[4-8] is an unambiguous determination of the presence of the Philadelphia chromosomal translocation. Since the specific amplification of a "fusion" sequence (e.g., BCR-ABL) is accomplished by using one primer complementary to the BCR sequence and the other to the ABL sequence, the PCR amplification of this unique translocation encoded fragment is a powerful and sensitive way to monitor minimal residual disease, capable of detecting one cell in 10^6 containing the fusion transcript. The role of the recently-identified class of tumor suppressor genes[9] is an intriguing and active area of investigation. PCR can contribute to the analysis of mutations in candidate loci[10] as well as to the detection of "allelic deletion"[11,12] by converting RFLP markers to PCR markers.

In addition to its role in the analysis of genetic disorders and cancers, PCR has proved extremely valuable in the diagnosis of a variety of infectious disease pathogens, as is discussed in some detail in Chapter 19. The detection of pathogens in environmental as well as in clinical samples has also been reported[13]. The analysis of pathogens (e.g., HPV[14]) and of oncogene mutations[15] in archival material like formalin-fixed paraffin-embedded tissue blocks makes possible retrospective molecular studies. The recently reported genetic diagnosis of phenylketonuria[16] and of cystic fibrosis carrier status[17] from old Guthrie cards (newborn blood spots stored on filters) also illustrates the capacity of PCR to analyze archival samples.

Although it is primarily the diagnosis of known pathogens and known genetic diseases and oncogene mutations that is the focus of this section, it is likely that PCR will also contribute to the identification of *new* pathogens, and possibly to understanding the viral etiology of some chronic diseases. The role of genetic alterations in known loci (e.g., the p53 gene) in neoplasia is being explored using PCR approaches[10] and the identification of new disease-related genes and mutations should emerge from the Human Genome project as well as from the work of many individual labs. Thus, PCR promises to play a critical research role in the identification of medically important sequences as well as an important diagnostic one in their detection.

REFERENCES

1. Feldman, G., Williamsen, R., Beaudet, A., and O'Brien, W. (1988) *Lancet* ii:102-103
2. Saiki, R., Walsh, P.S., Levenson, C.H., and Erlich, H.A. (1989) *Proc. Natl. Acad. Sci. USA*, in press.
3. Crescenzi, M., Seto, M., Herzig, G.P., Weiss, P.D., Griffith, R.C., and Korsmeyer, S.J. (1988) *Proc. Natl. Acad. Sci. USA* 85:4869-4873.
4. Kawasaki, E.S., Clark, S.S., Coyne, M.Y., Smith, S.D., Champlin, R., Witte, O.N., and McCormick, F.P. (1988) *Proc. Natl. Acad. Sci. USA* 5698-5702.
5. Lee, M.-S., Chang, K.-S., Freireich, E.J., Kantarjian, H.M., Talpaz, M., Trujillo, J.M., and Stass, S.A. (1988) *Blood* 72:893-897.
6. Price, C.M., Rassool, S., Shivji, M.K.K., Gow, J., Tew, C.J., Haworth, C., Goldman, J.M., and Wiedemann, L.M. (1988) *Blood* 72:1829-1832.
7. Dobrovic, A., Trainor, J., and Morley, A.A. (1988) *Blood* 72:2063-2065.
8. Hermans, A., Selleri, L., Gow, J., and Grosveld, G.C. (1988) *Blood* 72:2066-2069.
9. Knudson, A., Jr. (1985) *Cancer Res.* 45:1437-1447.
10. Baker, S.J., *et al.* (1989) *Science* 244:217-221.
11. Cavanee, W., *et al.* (1983) *Nature* 305:779-785.
12. Vogelstein, B., *et al.* (1989) *Science* 244:207-211.
13. Steffan, R.S., and Atlas, R.M. (1988) *Appl. Environ. Microbiol.* 54:2185-2191.
14. Shibata, D.K., Arnheim, N., and Martin, W.J. (1988) *J. Exp. Med.* 167:225-230.
15. Shibata, D.K., Martin, W.J., and Arnheim, N. (1988) *Cancer Res.* 48:4564-4566.
16. Lyonnet, S. *et al.* (1988) *Lancet* ii:507-508.
17. Williams, C., Weber, L., Williamson, R., and Hjelm, M. (1988) *Lancet* ii:693-694.

CHAPTER 14

Use of PCR in the Diagnosis of Monogenic Disease

Haig H. Kazazian, Jr.

INTRODUCTION

Since the fall of 1987, PCR technology has had a revolutionary impact upon the prenatal diagnosis of single gene disorders and carrier testing for these disorders. PCR technology has not yet expanded the repertoire of diseases which can be detected, but it has greatly expanded the options of the laboratory diagnostician. At Johns Hopkins it has allowed us to diagnose disorders with greater speed, greater accuracy, and with greater technical flexibility. Examples illustrating each of these improvements follow. Before October 1987, we carried out prenatal diagnosis of sickle cell anemia by Southern blotting for the mutation which usually required two or more weeks from the date of fetal sampling. Since October 1987, we have carried out these diagnoses by PCR techniques in two to four days from fetal sampling. Prior to October 1987, nearly all our prenatal diagnoses of ß-thalassemia were accomplished through indirect detection via linked DNA polymorphisms in the ß-globin cluster. Again, this work usually took two to four weeks to accomplish. Since October 1987, all our prenatal diagnoses of ß-thalassemia have been carried out by direct detection of the disease-producing mutations after PCR amplification of regions of the ß-globin gene. These methods provide increased accuracy and diagnosis usually within one week. Improved technical flexibility has meant that 1) if we are unable to determine a ß-thalassemia mutation in one parent or another, we can study DNA polymorphisms in the ß-globin cluster in a day or two, or 2) we can determine the extent of maternal contamination quickly using sequence differences between mother and fetus. These are just some examples of what PCR has meant to the gene diagnostic enterprise.

I will now cover a number of applications of PCR in gene diagnosis using specific examples. I will concentrate on use of PCR in 1) detection of known point mutations by dot-blot hybridization or restriction endonuclease digestion of the amplification product, 2) detection of known and unknown mutations by direct sequencing of the amplified product, 3) detection of suspected deletions by altered size of the PCR product or failure of specific amplification in multiplex reactions, and 4) detection of DNA polymorphisms for indirect detection of disease-producing mutations. I will also try to discuss technical difficulties, important controls, and limitations of the PCR technique as it is practiced in late 1988.

BACKGROUND ON GENE DIAGNOSIS IN PRENATAL DIAGNOSIS

A short background of gene diagnosis in prenatal diagnosis is in order. In 1978, Kan and Dozy demonstrated that a DNA polymorphism linked to the ß-globin gene could often be used as a marker to carry out indirect detection of sickle cell anemia.[1] This type of detection required family studies to determine in each parent the allele type of the polymorphism that is tracking with the normal ß-globin gene and the allele type that is tracking with the sickle ß-globin gene. Family studies may involve children of the couple or, if no children exist, the four parents of the couple. By 1982, it became possible to carry out prenatal diagnosis of sickle cell anemia by direct detection of the mutation because a restriction enzyme, *Mst*II, or its isoschizomers, *Cvn*I or *Oxa*NI, fails to cut the sickle ß-globin gene at the mutation site but does cleave the normal ß^A-globin gene at this site (Figure 1).[2-4] This advance meant that prenatal diagnosis of sickle cell anemia could be done more accurately and without the need for family studies. It is this kind of advance, the ability to detect disease-producing mutation directly and simply, that we seek in all gene diagnostic studies.

Initial efforts at diagnosis of any single gene disorder usually rely upon detection of the disorder using linked DNA polymorphisms and family studies. In late 1988, we are at this stage in the diagnosis of some cases of Duchenne muscular dystrophy, most cases of hemophilia A and hemophilia B, all cases of cystic fibrosis, Huntington disease, neurofibromatosis, and adult onset polycystic kidney disease. We have passed this preliminary stage to direct detection of the disease-producing mutations in sickle cell anemia (as mentioned above), ß-thalassemia,[5,6] α-

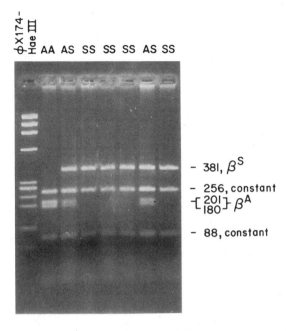

Figure 1. Ethidium bromide stain of *Cvn*I digested amplified DNA product from individuals with sickle cell trait (AS), normal ß-globin genes (AA), and sickle cell anemia (SS). Staining is carried out after electrophoresis in a 3% NuSieve agarose, 1% agarose mini-gel.

thalassemia,[7] most cases of Duchenne muscular dystrophy,[8-10] and some cases of hemophilia A[11,12] and retinoblastoma.[13]   In nearly all of these examples, because the sequences of critical DNA regions are known, PCR has become a valuable asset in prenatal diagnosis, presymptomatic diagnosis, and carrier detection.

USE OF PCR IN DIRECT DETECTION OF POINT MUTATIONS

Three techniques which are used following PCR in the diagnosis of point mutations, dot-blot hybridization,[14] restriction analysis,[15,16] and direct sequencing,[17-19] can best be discussed in the context of their use in diagnosis of ß-globin disorders.   Diagnosis of sickle cell anemia is carried out at Johns Hopkins by: 1) boiling fetal cells obtained by chorion villus sampling (CVS) or amniocentesis in 2 M NaC1, 0.1 N NaOH, 2) PCR amplification of a 725-bp region at the 5' end of the ß-globin gene using 30 cycles of PCR and 120 second chain extension times at 72°, 3) digestion of the 725-bp amplification product with *Cvn*I, 4) NuSieve agarose electrophoresis of the digestion product, 5) ethidium bromide staining of the DNA fragments, and 6) detection of DNA fragments under UV light and photography of the band pattern (Figure 1). Detailed protocols are presented at the end of this chapter. The primers used in the reaction were chosen such that the amplification product contains two constant *Cvn*I sites. Thus, at least three bands are visualized in every digested sample. Each sickle ß-globin gene has a 381-bp signature while the ß^A-globin gene is demonstrated by the cleaved products of this fragment, 201-bp and 180-bp fragments. Both ß^s and ß^A genes always demonstrate 256-bp and 88-bp fragments. Because of the clarity and high information content of the results, we prefer this method, restriction analysis of the PCR product, to dot-blot hybridization using oligonucleotides specific for the ß^s-mutation. However, a major limitation (also a limitation of Southern blot analysis) is the failure of the method to detect a deletion of the ß-globin gene which includes the sixth codon, the site of the ß^s mutation. An individual heterozygous for such a deletion and a ß^s-globin gene has the same band pattern as an individual with sickle cell anemia (two ß^s-globin genes) in this analysis. In a patient in whom the diagnosis is in doubt, the true diagnosis can best be obtained by studying the patient's parents. If both parents are ß^{A}ß^s by this technique, the diagnosis in the patient is sickle cell anemia. If one parent is ß^A only, while the other parent is ß^{A}ß^s, the diagnosis in the patient is heterozygosity for the ß^s gene and a deletion involving the ß-globin gene.

Direct detection of ß-thalassemia mutations became theoretically possible after the mutations leading to the disease were characterized.[20] Key ethnic groups at risk for ß-thalassemia in their offspring are Mediterranean, Middle Easterners, Asian Indians, Chinese and Blacks. As of late 1988, over 60 ß-thalassemia mutations are known[20,21] and the alleles leading to well over 99% of the ß-thalassemia genes in the world are characterized. Remaining alleles will be rare or present in ethnic groups with small populations at risk. Because each affected ethnic group has its own battery of ß-thalassemia mutations and, in general, 4-6 alleles make up greater than 90% of ß-thalassemia genes in any particular ethnic group,[20] the job of detecting the disease-producing mutations in both members of an at-risk couple is simplified.

Often a childless couple is referred for testing because both members appear to have ß-thalassemia trait by screening tests of their red cell volume and hemoglobin A$_2$ concentration. The approach for each couple is to test for the presence of alleles common in the ethnic group of the couple. Since September 1987, we have carried out 80 consecutive prenatal ß-thalassemia in which a large variety of ethnic groups were represented without resort to Southern analysis of DNA polymorphisms. All of these diagnoses were done by direct mutation detection in the couple and their fetus. This testing usually requires a combination of dot-blot hybridization and restriction analysis. Rarely, genomic sequencing is used. About half of the over 60 ß-thalassemia alleles can be detected by restriction endonuclease analysis of the amplified product (see Table 1). For Mediterranean couples, the nonsense codon 39, frameshift 6, IVS-2, nt 1, IVS-2, nt 745, IVS-1, nt 6, and -87 alleles are usually detected most easily by this method. The remaining common mutations in Mediterranean, IVS-1, nt 110, IVS-1, nt 1, and frameshift 8 are all detected by dot-blot analysis.

Table 1.

β-thalassemia mutation	Ethnic group	Affected restriction endonuclease site
Nonsense codon 39	Mediterranean	gains *Mae*I site
IVS-1, nt 6	Mediterranean	gains *Sfa*NI site (also polymorphic *Sfa*NI site at IVS-2, nt 16)
IVS-1, nt 1 (G-A)	Mediterranean	loses *Bsp*MI site
Frameshift 6	Mediterranean	loses *Cvn*I site
IVS-2, nt 745	Mediterranean	gains *Rsa*I site
-87	Mediterranean	loses *Avr*II site
IVS-2, nt 1	Mediterranean	loses *Hph*I site
βᴱ	Chinese	loses *Mnl*I site
Nonsense codon 17	Chinese	gains *Mae*I site
-29	Chinese, Black	gains *Nla*III site
Nonsense codon 43	Chinese	loses *Hinf*I site
-88	Black, Asian Indian	gains *Fok*I site
IVS-1, nt 1 (G-T)	Asian Indian	loses *Bsp*MI site
IVS-1, nt 5 (G-A)	Algerian	gains *Eco*RV site

In dot-blot analysis, 5-10% of the amplified product (usually a 725-bp fragment from the 5' one-half of the ß-globin gene) is dotted twice on a nitrocellulose filter. Two dots are usually tested from each individual for control purposes; to visualize the same positive or negative result with each dot. The dots are hybridized with a ^{32}P oligonucleotide (19-mer or 20-mer) or to a non-radioactively labeled probe[6] whose sequence is specific for the mutation under study. The same dots are then hybridized with a normal ß-globin sequence to determine that sufficient DNA was dotted to produce a positive signal. If the assay is negative with a mutant probe and positive with the normal probe, we know that the patient does not carry the particular mutation being analyzed. If both probes give positive results, the patient is heterozygous for the mutation under study. If the mutant probe gives a positive result and the normal probe does not, the patient is homozygous for the analyzed mutation (see Figure 2).

Rarely, the ß-thalassemia allele is still unknown in one member of a couple after DNA of both members of the couple has been analyzed for the common alleles present in their ethnic group. At that point key regions of the ß-globin gene are sequenced to determine the unknown disease-producing mutation.[17-19] Sequence analysis is carried out using T₇ DNA polymerase (Sequenase) on PCR-amplified regions of genomic DNA. These regions are: 1) the promoter, exon 1, intron 1, exon 2, and the 5' ninety nucleotides of intron 2 and 2) the 3' three hundred nucleotides of intron 2 and exon 3. This procedure yields the unknown mutation in nearly all cases studied. However, in at least two instances a mutation has not been found in the ß-globin gene or key 5' and 3' regions in carriers of mild ß-thalassemia alleles (unpublished data). In our hands, seven previously undescribed alleles have been found by direct sequence analysis of PCR-

Gene Amplification/Dot Blot Hybridization

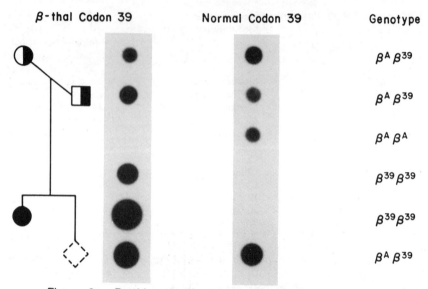

β-thal Codon 39   Normal Codon 39   Genotype

$\beta^A \beta^{39}$

$\beta^A \beta^{39}$

$\beta^A \beta^A$

$\beta^{39} \beta^{39}$

$\beta^{39} \beta^{39}$

$\beta^A \beta^{39}$

Figure 2.   Dot-blot showing the use of ASO probes on amplified ß-globin DNA for prenatal diagnosis of ß-thalassemia.   Both parents in the pedigree at the left carry the nonsense codon 39 mutation as demonstrated by hybridization of their amplified DNAs to the mutant ASO. Control samples, essential for monitoring the specificity of the wash, are in the third and fourth dots from the top and represent amplified DNA from individuals homozygous for normal ß-globin alleles and homozygous for the nonsense codon 39 ß-thalassemia allele, respectively.   Amplified DNA from this couple's affected child hybridizes to mutant ASO only and from the fetus to both the mutant and normal ASOs, which demonstrates that the fetus is a carrier of ß-thalassemia.

amplified ß-globin genes.[17,20]   Huisman and colleagues have found a comparable number of additional new alleles in populations different from those that we have studied.[22]

DETECTION OF DELETIONS BY PCR

Deletions of moderate size (under 1-1.5 kb) can be detected through use of primers located 5' and 3' to the deletion.   An example of such an analysis is seen in Figure 3.   A 619-bp deletion in the ß-globin gene is a common ß-thalassemia allele among Asian Indians.[22]   We detect this deletion using primers that yield a 1215-bp fragment in normal DNA and a 596-bp fragment when the 619-bp deletion is present.   The presence of both 1215-bp and 596-bp fragments after PCR of genomic DNA signifies heterozygosity for the deletion.

Alternatively, deletions may be detected by absence of a PCR product from a multiplex reaction in which other PCR products are detected.   Such an approach was first demonstrated by Chehab *et al.* for deletions involving the α-globin loci leading to α-thalassemia.[16]   These deletions were detected by absence of the α-globin specific product from PCR reactions containing both α-globin and ß-globin primers.   The presence of the ß-globin product in the absence of the α-globin product indicates homozygous deletion of the region containing at least one of the α-globin primers.

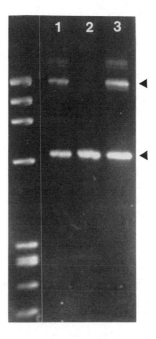

Figure 3. PCR analysis of a 619-bp deletion in the ß-globin gene.   This deletion is common in Asian Indians and eliminates the 3' end of IVS-2 and exon 3. A 1215-bp region is amplified with primers 5' and 3' to the deletion. *Hae*III digested 0/X174 fragments in the far left lanes are 1.35, 1.08, .87, .60, .31, .28, .27, .23 and .19 kb from top to bottom. In lanes 1 and 3 are DNA of parents heterozygous for the deletion and lane 2 is DNA of their child who is homozygous for the deletion. Arrows point to 1215- and 596-bp fragments. The origin of the PCR product larger than 1215 bp is unknown.

This type of approach has been expanded recently by Chamberlain *et al.* in studies of the dystrophin locus in males affected with Duchenne muscular dystrophy (DMD).[23]   DMD is an X-linked form of severe, early onset, muscular dystrophy due to gene defects in the dystrophin gene, which is approximately 2000 Kb in size.   Sixty to seventy percent of DMD alleles are deletions involving numerous regions of the dystrophin gene.[24] Chamberlain *et al.* have sequenced certain key regions and made primers specific for these and for others whose sequence was already known.   They then worked out conditions to carry out PCR with 9 primer pairs to amplify 9 different regions of the dystrophin locus in which 80-90% of the previously described deletions are located.   Using these primer pairs, they were able to identify many deletions (see Chapter 15 by R. Gibbs) which were then confirmed by Southern blot analysis using dystrophin cDNA probes.   This procedure of multiplex PCR for deletion analysis provides a rapid screening test which should detect the mutation in 50% of affected males.   The remaining families will still require cDNA probe analysis to detect other deletions and DNA polymorphism analysis to track the affected dystrophin locus within the family.

A couple of limitations of the multiplex PCR approach to detect deletions exist.   First, whenever one is studying the absence of a fragment one worries that the region which failed to amplify is actually present in genomic DNA but the amplification of that region failed for some other unknown reason.   We have seen examples in our lab of multiplex amplifications in which 3 of 3 primer pairs produced adequate PCR products in some normal samples while only 2 of 3 sets worked in others. Thus, one must be very careful in interpreting results in which 8 of 9 or 7 of 9 primer pairs produce an amplification product.   This result suggesting a gene deletion should be verified by repeated assays.   Moreover, in the early stages of clinical use of multiplex PCR, a presumed deletion detected by the technique should be verified by Southern blot analysis using the appropriate dystrophin cDNA probe.

A second major limitation is the inability of this technique to detect female carriers of deletions.   Such carriers will demonstrate the same band patterns as normal individuals.   This limitation is important because much of the laboratory analysis in families at risk for DMD involves carrier detection in families and carrier testing can now be successfully accomplished by quantitative Southern blot analysis using dystrophin cDNA probes.[9]

USE OF PCR FOR DETECTION OF LINKED DNA POLYMORPHISMS

Most gene diagnosis in prenatal diagnosis, carrier detection, and presymptomatic diagnosis is still carried out in situations where the gene of interest has not yet been isolated.   This is still the case in December, 1988 for cystic fibrosis, Huntington disease, and neurofibromatosis among others. Gene diagnosis in these instances depends entirely upon tracking the disease gene in affected families using DNA polymorphisms (RFLPs) linked to the disease locus. When DNA sequences surrounding important DNA polymorphisms become known, the presence or absence of a critical restriction endonuclease site can be assayed easily using PCR.

PCR analysis of DNA polymorphisms was first reported by Mullis and Faloona for sickle cell anemia[25] and by Kogan *et al.* for Hemophilia A diagnosis.[15] That report pointed out a potential limitation of PCR analysis in that sequences around one of the polymorphic sites, an *Xba*I site, are duplicated at another location in the genome. No other *Xba*I site is present in the short amplified region and the primers chosen amplify both the region of interest in the factor VIII:C gene and the second region. Thus, a male with the factor VIII:C *Xba*I site demonstrates the uncut PCR fragment derived from the second region in addition to the two smaller fragments derived from cleavage of the polymorphic *Xba*I site (Figure 4). Females who are +/- for the polymorphic site cannot be differentiated from those who are +/+. This complication does not exist when Southern blotting for the *Xba*I polymorphism is carried out. In Southern analysis, the fragments assayed are larger than the duplicated region resulting in an *Xba*I fragment from the factor VIII:C gene which differs in size from that of the second region (Figure 4).

Use of PCR for analysis of DNA polymorphisms has been reported for prenatal diagnosis of cystic fibrosis[26,27] and for presymptomatic and prenatal diagnosis of Huntington disease.[28] In addition, our laboratory is able to carry out ß-globin cluster haplotyping by PCR analysis of five restriction site polymorphisms (unpublished data). If we ever encounter difficulty in direct detection of sickle cell anemia or ß-thalassemia, haplotype analysis of various family members by PCR will allow rapid prenatal diagnosis by indirect detection. Thus, PCR has been used for diagnosis of many of the major monogenic disorders presently detected clinically: sickle cell anemia, the thalassemias, hemophilia A, Duchenne muscular dystrophy, cystic fibrosis, and Huntington disease.

PITFALLS OF THE TECHNOLOGY

A number of pitfalls have been mentioned earlier in this chapter. One primer pair may fail to amplify even though other primer pairs successfully amplify in the same DNA sample. These results are particularly distressing for the accuracy of deletion analysis using multiplex PCR reaction.

Contamination of buffers or PCR-primers with genomic DNA can lead to error and/or great frustration after one observes the same result in all samples amplified using the contaminated primer. Great care must be taken in pipetting, and in keeping plasticware and glassware used in PCR work separate from general labware.

Occasionally, for unknown reasons a sample of genomic DNA may fail to amplify. Often times, this is due to contaminants carried over from the DNA isolation procedure. This problem can usually be overcome by using less genomic DNA than usual and thus diluting the contaminants.

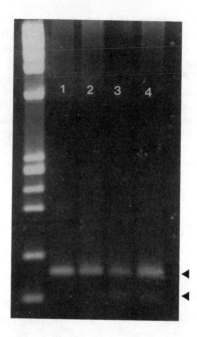

Figure 4A. PCR analysis of *Xba*I polymorphism in intron 22 of the factor VIII:C gene. On the far left is the *Hae*III digested ØX 174 marker (see Figure 3). When the polymorphic *Xba*I site is present, a 96-bp PCR fragment is cleaved to 68-bp and 28-bp fragments. The 28-bp fragment is not seen. A duplicate 96-bp region that does not contain a polymorphic *Xba*I site is always detected in this assay. Lanes are 1) male known (-) from Southern blot; 2) affected male fetus (-); 3) unaffected male fetus (+); 4) +/- female control, known +/- from Southern blot.

Amplification may also be obtained by titration of the magnesium concentration in the PCR reaction. Occasional DNA samples that are not degraded will still fail to amplify requiring the laboratory to obtain new tissue material for reanalysis.

FUTURE EXPECTATIONS

We predict that PCR will be used in carrier screening programs. Specifically, the best candidate diseases will be those with high incidence in one or more population groups, significant morbidity for affected individuals, and those for which there is knowledge about the gene of interest and the mutations that account for essentially all of the mutant alleles. Carrier screening for ß-thalassemia and sickle cell anemia by DNA

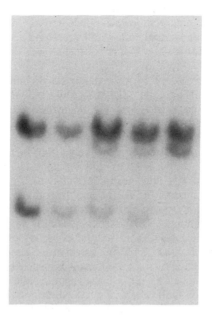

Figure 4B. Southern blot analysis of the *Xba*I polymorphisms in intron 22 of the factor VIII:C gene. The polymorphic fragments, 6.2-kb and 4.8-kb, are shown by arrows. A fragment, 6.6-kb, is present in all individuals after hybridization with an intron 22 genomic probe. Lanes from left are obligate carrier female (+/+), her unaffected son (+), three females from another family (+/-, +/-, -/-, respectively).

methods is possible but non-DNA tests are still cheaper and simpler than any gene diagnosis test.

Carrier screening for Tay-Sachs alleles in the adult Ashkenazi Jewish population by PCR has strong possibilities. Recent data suggest that two alleles producing defects in the α-chain gene of hexosaminidase A may account for essentially all Tay-Sachs genes in the Ashkenazi Jewish population.[30,31] If this is so, then screening for carriers of infantile Tay-Sachs disease genes might best be carried out by PCR analysis.

The alleles accounting for 50% of phenylketonuria (PKU) genes in the Northern European population have been characterized.[32,33] No other practical means for detecting carriers of PKU other than gene analysis exists. Thus, it is probable that some individuals at higher risk of being carriers because of their family histories will desire carrier screening when the alleles producing PKU are completely characterized.

When the CF gene is discovered and alleles producing CF are characterized, this disorder will be a prime candidate for carrier screening by PCR techniques. Carrier screening by gene analysis will be practical if the number of alleles is small (fewer than 10) and no simple biochemical test for carrier status emerges from knowledge of the defective protein. The possibility exists that population-based screening for carriers of the CF gene using PCR could have a major influence on the incidence of this disease.

The possibilities for use of PCR in diagnosis of monogenic diseases are exceedingly great. PCR-based diagnosis by analysis of linked DNA polymorphisms could be carried out for neurofibromatosis, retinoblastoma, adult onset polycystic kidney disease, and myotonic dystrophy, among others. PCR-based analysis of exon sequences by density-gradient gel electrophoresis in various disease genes, e.g., the factor VIII:C exons in hemophilia A patients, could lead to direct diagnosis of mutant alleles in most affected individuals.[34] New techniques in which oligonucleotides with poly-T tails are immobilized on filters and hybridized with a biotinylated PCR fragment of the region of interest from patient's DNA may revolutionize gene diagnosis.[35] This technique will simplify assay for 5-10 or more alleles in the genome of an individual as is now carried out for ß-thalassemia. The technique may be used in a general genetic profile analysis whereby the genome of an individual is tested for the presence of a number of deleterious genes, e.g., ß-thalassemia alleles pertinent to his ethnic group, the sickle cell allele. PKU alleles, CF alleles, Tay-Sachs alleles, etc. A multiplex PCR might be done in order to screen a number of genes simultaneously.

The possibility of testing for genetic traits in the preimplantation embryo arises because of the availability of PCR. One or two cells could be removed from the embryo at the blastomere stage and using PCR techniques a genetic diagnosis could be made on the DNA of these cells. The embryo could undergo further growth *in vitro* and later be implanted into the uterus. In other species, apparently normal fetal development follows removal of a small number of cells from the blastomere followed by reimplantation. Thus, there is a practical possibility for genetic diagnosis of the preimplantation embryo. However, the ethics of this procedure would be a matter of significant debate, and it is unlikely that an experimental review board will allow implantation of a human embryo after it has undergone removal of a small number of cells at the blastomere stage.

SUMMARY

Use of PCR for diagnosis of monogenic diseases has become well established. In 1988 at Johns Hopkins, all prenatal diagnoses for sickle cell anemia and ß-thalassemia and one-half of diagnoses for hemophilia A were carried out using PCR techniques. Overall, about one-half of the approximately 440 families studied by us in 1988 received diagnostic information obtained following PCR. In 1989 at Johns Hopkins, one-half of all DMD diagnoses, one-half or more of CF diagnoses, and one-half or more of Tay-Sachs screening will be carried out using PCR techniques.

The use of Southern blotting in this diagnostic work will not be eliminated, but we expect that the use of Southern blots in 1989 will decline to about 30-40% of its use in 1987. PCR techniques will probably replace about 90% of Southern blot use in diagnosis of monogenic diseases within 3-5 years. However, care will have to be taken to maintain the highest diagnostic accuracy during this transition period.

PROTOCOLS

Making Stock Solutions

For PCR:

40 mM deoxynucleotides (10 mM each dATP, dGTP, dCTP, dTTP); dilute with distilled H_2O; store at -20°C; thaw only for immediate use.

10 X reaction buffer:　　　　　500 mM KCl
　　　　　　　　　　　　　　100 mM Tris.HCl, pH 8.3
　　　　　　　　　　　　　　15 mM $MgCl_2$
　　　　　　　　　　　　　　0.1% gelatin

Store at -20°C

Primers diluted to 10 μM. Primer molarities calculated as:

　　　　　　　　　　363 (# of Gs) + 323 (# of Cs) + 347 (# of As) + 322 (# of Ts) = grams/mole

Store at -20°C

For electrophoresis:

20 X electrophoresis buffer:　　0.8 M Tris
　　　　　　　　　　　　　　0.26 M NaAcetate　pH 8.0
　　　　　　　　　　　　　　40 mM Na_2EDTA

Bromphenol blue dye:　　　　10% ficol
　　　　　　　　　　　　　　10% glycerol
　　　　　　　　　　　　　　0.2% BPB

Ethidium bromide:　　　　　10 mg/ml in distilled H_2O

For dot-blot:

Denaturing solution:　　　　0.4 N NaOH, 25 mM Na_2EDTA

20 X SSPE:	3.6 M NaCl	
	20 mM NaH_2PO_4	pH 7.4 with
	20 mM Na_2EDTA	NaOH to get into solution
50 X Denhardt's:	1% Ficoll	
	1% PVP (polyvinyl pyrrolidone)	
	1% BSA	
Prehybridization buffer:	5 X SSPE	
	0.5% SDS	
	5 X Denhardt's	
Wash buffer:	2 X SSPE	
	0.1% SDS	

For end-labeling:

Allele-specific oligonucleotide (ASO)
Kinase buffer recommended by T4 kinase supplier
Biogel P4 swollen in distilled H_2O

Preparation of Fetal DNA for PCR

Amniotic fluid:

- Take at least 1.5 cc of fluid and spin for 5 min. (Can use more fluid but doesn't seem to make a difference.)
- Discard supernatant. Wash pellet with TE buffer (10 mM Tris.HCl pH 7.4, 1 mM Na_2 EDTA).
- To washed cell pellet add 10 λ of .1 M NaOH 2 M NaCl and vortex.
- Boil at 100°C for 2 min. Re-vortex.
- Spin for 10 min. Transfer supernatant.
- Take an aliquot of supernatant and dilute 1:10 with distilled H_2O and then use 4 λ of diluted sample in 100 l PCR reaction.

CVS and cultured cells:

- Wash villi with TE buffer x 2.
- To washed villi add 20-30 λ (depending on amount of villi) of .1 M NaOH, 2 M NaCl at 67°C (will have to heat solution before using).
- Vortex.
- Boil at 100°C for 2 min. Re-vortex.
- Spin 10 min. Transfer supernatant.
- Take an aliquot of supernatant and dilute 1:10 with distilled H_2O and then use 4 λ of diluted sample in PCR reaction.

Amplification and Analysis

PCR reaction:

In 0.5 ml microcentrifuge tube, combine:
- 50 ng - 1 μg DNA
- 2 μl of 40 mM dNTPs

- 10 μl of 10 X reaction buffer
- 5 μl of each 10 μM primer
- 2.5 U *Taq* polymerase (with each set of amplification reactions, you can make a "stock" dilution of enzyme in 1 X reaction buffer for pipeting ease)
- Distilled H_2O to total volume of 100 μl
- Mix and spin down in microfuge; overlay with mineral oil

Reaction: 30 cycles
    1) 6 min 94°C . . . . time delay file
    2) (30 sec 94°C, 30 sec 55°C, 2 min 72°C) (times and temperatures vary with primers used) X 25-30 cycles . . . . step cycle file
    3) 10 min 72°C . . . . time delay file
    4) 4°C . . . . soak file

Electrophoresis (to check amplification product size and quantity):

3% NuSieve agarose, 1% agarose mini-gel.
In one lane, load ~1-2 μg of ØX174 DNA digested with *Hae*III (size marker) with BPB dye.
For each PCR sample, load 10 μl of product with 1 μl of marker dye.
Electrophoresis at 100 volts until BPB dye has migrated about 5 cm.
Stain gel in EtBr (diluted to 50 μg/25 ml) for 5-10 min; visualize on UV box (302 nm) and photograph.

Digestion for detection of β^s-globin allele:

Mix 10 μl of PCR product with 10 U *Cvn*I restriction endonuclease (BRL) in 20 μl total volume under conditions recommended by supplier of *Cvn*I.
30°C overnight digestion.
Electrophoresis in 3% NuSieve agarose, 1% agarose gel containing ØX174 marker until BPB has migrated approximately 7 cm.
Stain in EtBr as above; visualize on UV box and photograph.

Dot-blotting:

Denature 10 μl per dot (or more if amplification was poor) of PCR product in 100 μl .4 N NaOH, 25 mM Na_2EDTA and vortex briefly.
Wet Gentran filter in distilled H_2O and place on vacuum driven dot-blot apparatus.
Load 100 μl of sample/well. It's a good idea to dot each sample twice.
Each filter also needs samples from positive and negative controls for each probe that will be used on it. (You can re-use filters after soaking filter in denaturing buffer for 5-10 min and rinsing with a neutralizing buffer 2 times.)
Turn vacuum on until liquid has been evacuated from well. Turn vacuum off.
Rinse each well that was used with 200 μl 20 X SSPE and put 100 μl dilute BPB dye in some of the unused wells for filter orientation.
Apply vacuum making sure liquid is pulled out of each well. Keep vacuum on about 2 min.
Air dry filter.
Expose DNA side of filter to 254 nm UV source for 15 sec or 302 nm UV source for 2 min. Bake briefly to dry filter at 65-80°C and then label filter with marking pen.

* Prehybridize each filter with 5 ml of prehybridization buffer at 30°C for 15-30 min.
Add end-labeled ASO (about 5×10^6 cpm/ml of prehybridization buffer).
Hybridize at approximately the Tm of ASO less 2°C. Tm is calculated as:
4(# G + # C) + 2(# A + # T).
Wash in 2 X SSPE, 0.1% SDS   1) room temperature x 2.... just rinse
                                             2) 10 min wash at Tm of ASO in shaking
                                                 water bath
Blot wet filter, then put filter between plastic wrap. Expose to X-ray film and light screens at -80°C for 30 min.
Develop film.

* It is often sufficient to simply wet the filter with prewarmed buffer and proceed directly with the hybridization.

REFERENCES

1. Kan, Y.W., and Dozy, A.M. (1978) *Proc. Natl. Acad. Sci. USA* 75:5631.
2. Chang, J.C., and Kan, Y.W. (1982) *N. Engl. J. Med.* 307:30.
3. Orkin, S.H., Little, P.F.R., Kazazian, H.H., Jr., and Boehm, C.D. (1982) N. Engl. J. Med. 307:32.
4. Wilson, J.T., Milner, P.F., Summer, M.E., Nallaseth, F.S., Fadel, H.E., et al. (1982) *Proc. Natl. Acad. Sci. USA* 79:3628.
5. Pirastu, M., Kan, Y.W., Cao, A., Conner, B.J., Teplitz, R.L., and Wallace, R.B. (1983) *N. Engl. J. Med.* 309:284.
6. Saiki, R.K., Chang, C.-A., Levenson, C.H., Warren, T.C., Boehm, C.D., Kazazian, Jr., H.H., and Erlich, H.A. (1988) *New Engl. J. Med.* 319:537.
7. Embury, S.H., Lebo, R.V., Dozy, A.M., and Kan, Y.W. (1979) *J. Clin. Inv.* 63:1307.
8. denDunnen, J.T., Bakker, E., Klein Breteler, E.G., Pearson, P.L., and vanOmmen, G.J.B. (1987) *Nature* 329:640.
9. Forrest, S.M., Cross, G.S., Flint, T., Speer, A., Robson, K.J.H., and Davies, K.E. (1988) *Genomics* 2:109.
10. Darras, B.T., Blattner, P., Harper, J.F., Spiro, A.J., Alter, S., and Francke, U. (1988) *Am. J. Hum. Genet.* 43:620.
11. Gitschier, J., Wood, W.I., Tuddenham, E.G.D., Shuman, M.A., Goralka, T.M., Chen, E.Y., and Lawn, R.M. (1985) *Nature* 315:427.
12. Antonarakis, S.E., and Kazazian, H.H., Jr. (1988) *Trends in Genet.* 4:233.
13. Friend, S.H., Bernards, R., Rogel, J., Weinberg, R.A., Rapaport, J.M., Albert, D.M., and Dryja, T.P. (1986) *Nature* 323:643.
14. Saiki, R.K., Bugawan, T.L., Horn, G.T., Mullis, K.B., and Erlich, H.A. (1987) *Nature* 324:163.
15. Kogan, S.C., Doherty, M., and Gitschier, J. (1987) *N. Engl. J. Med.* 317:985.
16. Chelab, F., Dogerty, M., Cai, S., Kan, Y.W., Cooper, S., and Rubin, E. (1987) *Nature* 329:293.
17. Wong, C., Dowling, E.E., Saiki, R.K., Higuchi, R., Erlich, H.A., and Kazazian, H.H., Jr. (1987) *Nature* 330:384.
18. Engelke, D.R., Honer, P.A., and Collins, F.S. (1988) *Proc. Natl. Acad. Sci. USA* 85:544.
19. Saiki, R.K., Gelfand, D.H., Stoffel, S., Scharf, S.J., Higuchi, R., Horn, G.T., Mullis, K.B., and Erlich, H.A. (1988) *Science* 239:489.

20. Kazazian, H.H., Jr., and Boehm, C.D. (1988) *Blood* 72:1107.
21. Gonzalez-Redondo, J.M., Stoming, T.A., Lanclos, K.D., Gy, U.C., Kutlar, A., Kutlar, F., Nalcatsuji, T., Deng, B., Han, I.S., McKie, V.C., and Huisman, T.J.H. (1988) *Blood* 72:1007.
22. Spritz, R.A., and Orkin, S.H. (1982) *Nucl. Acids Res.* 10:8025.
23. Chamberlain, J.S., Gibbs, R.A., Ranier, J.E., Nguyen, P.N., Farwell, N.J., and Caskey, C.T. *Nucl. Acids Res.*, in press.
24. Koenig, M., Hoffman, E.P., Bertelson, C.J., Monaco, A.P., Feener, C., and Kunkel, L.M. (1987) *Cell* 50:509.
25. Mullis, K.B., and Faloona, F. (1987) *Meth. in Enzymol.* 155:35.
26. Feldman, G.L., Williamson, R., Beaudet, A.L., and O'Brien, W.E. (1988) *Lancet* ii:102.
27. Feldman, G.L., O'Brien, W.E., Durtschi, B., Gardner, P., Williamson, R., and Beaudet, A.L. (1988) *Am. J. Hum. Genet.* 43:A83 (abstract).
28. Brock, D.J.H., Mcintosh, I., Curtis, A., and Millan, F.A. (1988) *Am. J. Hum. Genet.* 43:A79 (abstract).
29. Myerowitz, R. (1988) *Proc. Natl. Acad. Sci. USA* 85:3955.
30. Arpaia, E., Dumbrille-Ross, A., Maler, T., et al. (1988) *Nature* 333:85.
31. Myerowitz, R., and Costigan, F.C. (1988) *J. Biol. Chem.* 263:18587.
32. DiLella, A.G., Marvit, J., Lidsky, A.S., Guttler, F., and Woo, S.L.C. (1986) *Nature* 322:799.
33. DiLella, A.G., Marvit, J., Brayton, K., and Woo, S.L.C. (1987) *Nature* 327:333.
34. Sheffield, V.C., Cox, D.R., Lerman, S.L., and Myers, R.M. *Proc. Natl. Acad. Sci. USA*, in press.
35. Saiki, R.K., Walsh, P.S., Levenson, C.H., and Erlich, H.A. (1989) *Proc. Natl. Acad. Sci. USA*, in press.

CHAPTER 15

Diagnosis of New Mutation Diseases Using the Polymerase Chain Reaction

R.A. Gibbs, J.S. Chamberlain, and C. Thomas Caskey

INTRODUCTION

Duchenne muscular dystrophy (DMD) and the Lesch-Nyhan (LN) syndrome are human genetic diseases that arise from heterogeneous mutations at the dystrophin and the hypoxanthine phosphoribosyltransferase (HPRT) loci, respectively.[1,2] The genes that are deficient in the two diseases are located on the X-chromosome but have many other different features (Figure 1). While the DMD gene spans more than two megabases of DNA, is split into more than sixty exons and has a mature mRNA that is approximately 14 kb in length, the HPRT gene is just 44 kb in length[3] and produces a 1.6-kb mature mRNA. Dystrophin is expressed in muscle and brain and is not detectable in other tissues.[4,5] In contrast, the HPRT gene product is ubiquitously expressed.[6] Examination of the molecular basis of LN and DMD by Southern analysis reveals that the spectrum of mutations in the two disorders is also different. Approximately 60% of DMD mutations are associated with DNA deletions[7] while only 15% of LN patients have abnormal Southern patterns that indicate DNA deletions or other gross

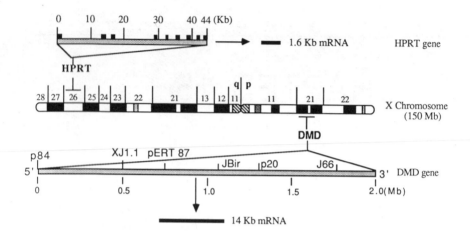

Figure 1. Organization of the genes defective in Duchenne muscular dystrophy (dystrophin) and the Lesch-Nyhan syndrome (HPRT). The large size of dystrophin and the predominance of DNA deletions that occur at this locus contrasts with the smaller HPRT gene that is frequently inactivated by point mutation.

rearrangements.[8]   Approximately 80% of LN cases also produce HPRT mRNA that appears normal by Northern analysis, suggesting a preponderance of single DNA base changes.[9]   As an additional difference, several informative DNA sequence polymorphisms have been identified within the dystrophin gene[7,10] while there are few genetic markers that are both informative and closely linked to HPRT.

The diverse genetic organization represented by these two loci provides for an excellent model for mutation detection that can be applied to most genetic diseases.   The development of a unified strategy for simultaneous identification of molecular lesions as varied as gross DNA rearrangements and single DNA base substitutions represents a considerable challenge that would have seemed unattainable until a very short time ago.   It is now possible to use polymerase chain reaction (PCR)[11-14] based techniques to perform DNA deletion screening and to enable the detection of single base alterations.   This report describes the use of PCR to enable detection of DNA deletions in the DMD gene, manual and automated direct DNA sequence analysis of mutant HPRT alleles, and simplified detection of DNA variations of known sequence at both loci.   Together these strategies form the basis of our currently favored approach for molecular diagnosis of all possible gene mutations (Figure 2).

PCR STRATEGIES FOR DIAGNOSIS OF NEW MUTATION DISEASES

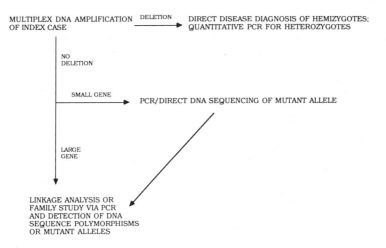

Figure 2. A general strategy for the molecular diagnosis of newly arising human disease mutations. Multiplex DNA amplification can be used as a primary screen to detect DNA deletions. Cases not able to be diagnosed by the deletion studies can be further studied *via* DNA linkage analysis or, if the target fragments are small enough, by direct sequence analysis of the mutations. The mutant sequence information can then be used in simplified assays to diagnose future disease cases in the family or to provide carrier diagnosis.

IDENTIFICATION OF NEW MUTATIONS AT THE HPRT GENE LOCUS

The frequency of LN is approximately one in one hundred thousand live births[1] and, consequently, the widespread screening for the disorder is not currently a practical objective. DNA-based tests are usually requested for diagnostic confirmation when an affected individual is identified by clinical and biochemical criteria and when other family members may be at risk as carriers. Until recently, the typical sequence of events would be for blood samples to be collected from all relevant family members to enable transformed cell lines to be established. DNA from the affected male would then be examined by Southern analysis using cloned cDNA probes in the hope of identifying a characteristic alteration in the normal banding pattern. A small fraction of families (<15%) are satisfied by this approach but, in the majority of cases, the cell lines must be subjected to drug selection and HPRT activity assays to determine the genetic status of each individual. Alternatively, an enzyme assay may be performed on hair root follicles. Although these methods can provide definitive diagnosis, they are each laborious, subjective, and have the major drawback that they cannot be quickly applied to prenatal diagnostic studies.

PCR now offers the opportunity to directly identify the mutant alleles in an affected male and then to use the DNA sequence information to design rapid assays to detect the alteration in other family members.   The new approaches take advantage of the compact size of the HPRT peptide coding sequence   and   the   presence   of   sufficient   HPRT   mRNA   in   most lymphoblastoid   cell   lines   derived   from   LN   patients   to   allow   the   PCR amplification of HPRT complementary DNA (cDNA).   The overall scheme is illustrated in Figure 3A and the experimental protocols described below (Protocols A-C).   First, cellular RNA is copied *via* reverse transcriptase to generate cDNA.   Specific oligonucleotide primers are then used to amplify a single fragment that contains the entire HPRT peptide coding region (Figure 3B).   DNA base alterations in the HPRT cDNA are identified by direct DNA sequencing of the amplified material using either manual or automated DNA sequencing techniques (Figures 3C and 4).

Direct sequencing is now the preferred method for analysis of all mutant HPRT   mRNA   that   can   be   PCR   amplified.   Previously,   we   used heteroduplex mapping techniques to localize individual mutations but this did not provide the precise DNA sequence of the alterations.[15,16]   The attraction of heteroduplex "scanning" methods is that relatively long fragments can be analyzed in a single assay.[17]   However, the direct DNA sequence analysis is sufficiently simple and rapid for the complete determination of all 651 bases coding for HPRT in order to define the mutation.

Several features of our protocols either deviate slightly from other reported methods or could be altered to accommodate different objectives. The rationale for each step is as follows:

1.   Starting material.   Cesium chloride gradient purified RNA[18] has always been used, but others have reported successful amplification of RNA isolated from crude cell extracts.[19]   The best choice of starting material will depend primarily on the number of samples to be analyzed and the abundance of available cells.   To assay a small number of established cell lines, the ultracentrifugation method is quite convenient.

2.   cDNA synthesis.   There are at least three methods for priming the synthesis of cDNA: random priming, oligo(dT) priming, and specific oligomer priming.   Priming with a specific oligomer has been avoided as the resulting cDNA cannot be used as a template for PCR of other DNA fragments.   In addition, the conditions for annealing of a specific oligomer must be stringently controlled.   There seems little difference between the performance of the non-specific priming methods although oligo(dT) has the theoretical disadvantage of less efficient coverage of the 5'-end of the message.   Thus, the random hexamers offer the advantages of a simple protocol that yields a product that can be used for amplification in multiple PCRs.   Note that controlled synthesis of a second cDNA strand is unnecessary.   However, including the alkaline hydrolysis step after the cDNA synthesis improves the quality of the final product as determined by agarose gel electrophoresis.

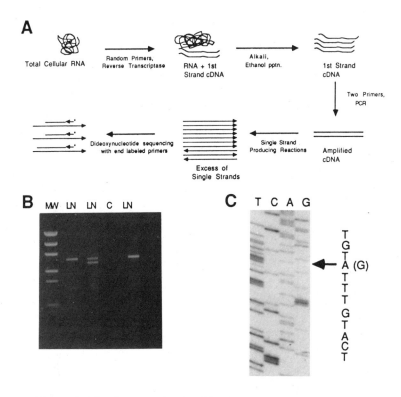

Figure 3. **A.** Strategy for the identification of new mutations in human HPRT cDNA. The details of the scheme are explained in the text. Briefly, RNA is copied *via* reverse transcriptase and the cDNA amplified using HPRT specific PCR primers. A second PCR-like reaction is performed using a single oligonucleotide primer and a small aliquot of the first reaction to generate a single-stranded mixture suitable for DNA sequencing. Dideoxynucleotide sequencing is carried out using end-labeled DNA sequencing primers. **B.** MW: *Hae*III digested φX174 DNA molecular weight standards. LN: Examples of PCR amplification of a 771-base fragment containing the entire human HPRT cDNA peptide coding region using RNA isolated from LN patients. Note the presence of two amplified bands in one lane, corresponding to alternative HPRT RNA splicing products in this patient (Gibbs *et al.*, manuscript in preparation). The specific oligonucleotides used for the PCR amplification were (sense) 5'-CTC TGC TCC GCC ACC GGC TTC CT-3' and (antisense) 5'-GGC GAT GTC AAT AGG ACT CCA GAT G -3' C: Control lane showing the absence of PCR product when reverse transcriptase is omitted from the cDNA synthesis reaction. **C.** Manual direct DNA sequence analysis showing the substitution of an adenosine residue in place of guanine in LN patient RJK 1727. See Protocol C.

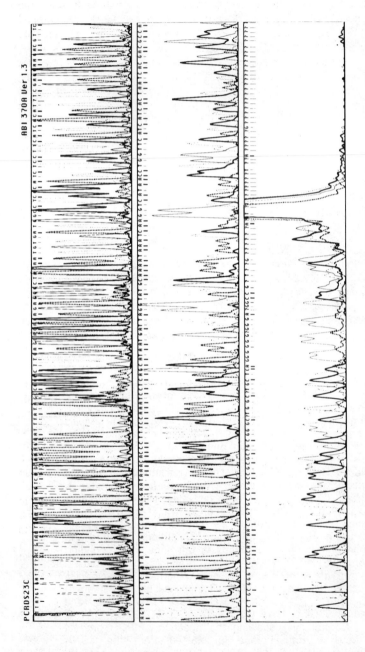

Figure 4. Identification of a single DNA base deletion in human HPRT cDNA by automated direct DNA sequencing, according to the method described in Protocol C. The unedited output from the computer is shown. The single DNA base deletion was independently confirmed by manual direct DNA sequencing and occurs twelve bases from the left in the second panel (normal sequence GTCCATAATTA; deleted base underlined). The peak of activity at the end of the third panel is the result of "run-off" labeling of the PCR fragment. "C" signal, solid thick lines; "G" signal, solid thin lines; "A" signal, broken thick lines; "G" signal, broken thin lines.

3. Amplification reactions. Many variations of the basic PCR have been described. Some of the variables must be explored each time the reaction system is established and other investigators recommend titration of mg^{++}, primer and deoxyribonucleotide triphosphate concentrations for optimal amplification of each individual DNA fragment. We have had success with the recipe described below that is based on the method of Kogan *et al.*[13] Multiple primer sets, including the pair described in Figure 5 and others complementary to the human HPRT cDNA,[20] have performed well when the reaction annealing temperatures and extension times have been optimized. We routinely use 23-25 base oligonucleotides for the cDNA amplifications, although 16- and 18-mers have functioned well in some circumstances. Generally, the longer oligomers are less inclined to generate non-specific amplification products at the optimum reaction stringency. Typically, a new primer set will be tested with 30 sec denaturation (93°C), 30 sec annealing (45°C), and extension at 65°C for 1 min for each 300 bases to be amplified. The product is assayed by agarose gel electrophoresis and the reaction repeated at a higher annealing temperature if fragments other than the expected species are observed. When the optimum annealing temperature is determined, the reaction extension times are reduced to determine a minimum period that does not affect the efficiency of the overall amplification. All primer pairs tested so far have performed well enough to generate at least 1.0 μg of amplified HPRT cDNA from the equivalent of less than 0.5 μg of total cellular RNA following 30 PCR cycles on an automated thermocycler.

4. Synthesis of single strands. A separate reaction is used to generate single DNA strands for DNA sequencing, which differs from the single unbalanced priming protocol described by Gyllensten and Erlich.[21] In our hands, the separation of the two steps has produced more reliable results as well as offering an opportunity to monitor the success of the primary amplification of the HPRT cDNA. This is valuable because, in some cases, the agarose gel banding pattern of the double-stranded products is abnormal and therefore diagnostic of the HPRT mutation (Figure 3B). Several DNA sequencing reactions are required to obtain complete coverage of the entire HPRT peptide coding region[20] and, therefore, the initial PCR can be used to seed multiple separate single strand producing reactions. There is no need to return to the cDNA synthesis products in each case. Increasing the distance between the primer used to generate single-stranded DNA template and the DNA sequencing primer can diminish the signal from the sequenced products; however, primers as far as 650 bases apart have functioned reliably.

5. Manual DNA sequencing. The salient features of the DNA sequencing reactions are: the use of the modified T7 DNA polymerase (Sequenase) in preference to *Taq*,[22] klenow, or reverse transcriptase; 5'-end labeled primers that are "internal" to the PCR primers; a reaction buffer that differs slightly from the generally recommended

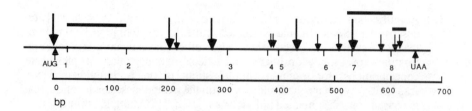

HPRT LOCUS

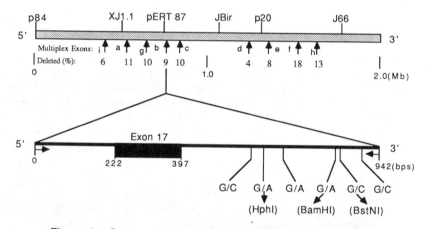

B.

DMD LOCUS

Figure 5. Summary of mutations identified in this study. **A.** HPRT mutations. A schematic representation of the human HPRT cDNA is shown with the position of single DNA base substitutions (small arrows), small deletions and insertions (large arrows), and exons involved in RNA splicing errors (horizontal bars) indicated above. Further explanation of each mutation is given in ref. 20, with the exception of a recently identified splice error in exon two. **B.** DMD mutations. TOP, Schematic illustration of the DMD gene illustrating the relative locations of the nine currently amplified regions (arrows) relative to six genomic probe markers. References for the genomic probes and a description of exons A-F are given in ref. 25. The other regions flank: i) exon 4; g) exon 12; h) the 3.1-kb *Hind*III fragment exon detectable with cDNA probe 8[7] (Chamberlain *et al.*, manuscript in preparation). Also indicated is the percentage of DMD patients carrying deletions of each of the nine regions.[7,37] BOTTOM, Expanded view of the region flanking exon 17, illustrating the location of 6 common polymorphisms useful for PCR-mediated haplotyping (see Figure 7; Chamberlain *et al.*, manuscript in preparation).

sequenase buffer; dideoxynucleotide stock mixes each 80:8 μM that are commercially available; and reaction temperatures of 50°C, instead of the usual 37°C. The reaction temperature has been a particularly important factor for the development of clear sequencing ladders. Manual sequencing reactions routinely yield a minimum of 200 bases and at least three oligonucleotide primers are usually required to enable complete coverage of the HPRT peptide coding region.

6. Automated DNA sequencing. The success of the direct sequencing with ^{32}P end-labeled DNA primers and a few negative experiences during attempts to obtain sequence by incorporation of radioactive dATP prompted our choice of the Applied Biosystems Inc. automated fluorescent DNA sequencer that employs end-labeled fluorescent DNA primers.[23,24] The manipulations for the automated DNA sequence analysis are essentially a scaling up of that for manual DNA sequencing. Generally, two single strand producing reactions are pooled before distribution of aliquots to be annealed to each of the custom produced primers. The remainder of the protocol is identical to the manual sequencing except that all four reactions are pooled prior to loading on the gel. Automated sequencing reactions routinely yield 275-350 bases of sequence and thus three separate oligonucleotide primers allow coverage of the HPRT coding sequence with generous overlap.

The sequence of altered HPRT cDNAs has been determined in sixteen independent cases of HPRT deficiency in our laboratory by these methods (Figure 5A). The cDNA sequence information provides the basis for design of simple assays for mutant alleles in other family members (see below). The application of further tests, however, assumes that the altered base sequences identified in the cDNA directly reflect alterations in the gene. The exceptions to this are those base changes that alter RNA processing, such as RNA splicing mutations. Currently, we estimate that approximately 15% of LN cases are in this category. Elucidation of the primary defects in these cases awaits the incorporation of DNA sequence information from regions that flank the individual HPRT exons[25] in order to design PCR primers to amplify genomic sequences.

DETECTION OF DELETIONS AT THE DUCHENNE MUSCULAR DYSTROPHY LOCUS

DMD is among the most common human genetic disorders, affecting approximately 1 in 3,500 male births, and one-third of all cases appear to arise *via* new mutation.[2] Southern analysis has been the method of choice to detect deletions at the dystrophin gene, while techniques to detect point mutations have not yet been successfully applied to this enormous locus. Unfortunately, there are several major limitations that prevent widespread and routine use of Southern analysis for DMD diagnosis. Exons of this gene are located on more than 65 genomic *Hin*dIII restriction fragments,

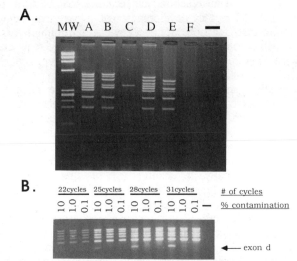

A.

MW A B C D E F —

B.

22cycles	25cycles	28cycles	31cycles	# of cycles

10 1.0 0.1 10 1.0 0.1 10 1.0 0.1 10 1.0 0.1 — % contamination

← exon d

Figure 6. Detection of DNA deletions at the DMD locus by multiplex DNA amplification. **A.** Demonstration of deletions in DMD males. M.W:*Hae*III digested φX174 DNA molecular weight standards. A-F: Amplification products obtained from patient DNA. G: Control reaction with no DNA template added. The analysis was performed as described in Protocol D. From top to bottom, the amplified fragments correspond to regions e, f, c, b, h, a, g, d and i, respectively (Figure 5B). **B.** Effect of maternal DNA contamination on multiplex deletion detection. A DNA sample deleted for region D (Figure 5B) was mixed with the indicated percentage of DNA from a normal female and amplified for the indicated number of cycles. The analysis was performed as described in Protocol D except that only 6 primer sets were included in the reaction (a-f). Following 25 reaction cycles, the sample containing 10% contamination has begun to reveal fragment D (reprinted from ref. 25 with permission).

which necessitates hybridization of blots with 7-9 separate dystrophin cDNA subclones to resolve each exon-containing restriction fragment for diagnosis of genomic alterations.[7,10] Furthermore, Southern analysis is an expensive, tedious, and time-consuming technique that requires the use of radioisotopes making it unsuitable for routine use in clinical laboratories.

The polymerase chain reaction offers advantages for the diagnosis of dystrophin gene abnormalities due to its ease of application, speed, and sensitivity. In addition, the ability to multiplex PCR, i.e., amplify many sequences simultaneously, has permitted the development of a rapid method capable of detecting 80 to 90% of all dystrophin gene deletions.[26] basis of the method relies on the observation that although dystrophin deletions arise in a heterogeneous manner, they are generally large (typic. several hundred kb) and are concentrated around two specific regions of the gene. Thus, a vast majority of DMD deletions is detectable by assaying for the presence or absence of only a small percentage of the total exons in the gene. Since the average intron size in this gene is approximately 35 kb, each exon must be amplified on a separate DNA fragment.

We are currently combining 18 separate oligonucleotide primers in a single reaction to achieve simultaneous amplification of 9 deletion prone exons of the gene (Figure 6). The assay is rapid (about 5 hr) and large quantities of premixed reactions can be prepared and stored at -70°C. For analysis, a tube is thawed, DNA and *Taq* polymerase are added, and the reaction is performed on an automatic thermocycler. The results are easily determined by electrophoresing a small aliquot of the reaction products on agarose gels and visually identifying which of the nine fragments have been amplified. A schematic representation of the DMD gene is shown in Figure 5B indicating the relative location of the nine currently amplified regions relative to a number of commonly used diagnostic probes. Figure 6A displays the results obtained *via* multiplex amplification from the DNA of six different DMD patients. To date, we have analyzed approximately 150 patients by this method and, in each case, the results have been subsequently confirmed *via* Southern analysis. In only two cases have all nine regions been deleted and we recommend such results be confirmed *via* Southern or dot-blot analysis with an appropriate cDNA probe. Alternatively, an additional set of primers derived from elsewhere in the genome could be added to the reactions to serve as a positive control for amplification.

Successful development of multiplex PCR for DMD diagnosis required both appropriate DNA sequence data for synthesis of primers and modification of reaction conditions to permit reproducible and specific co-amplification of each targeted DNA fragment. The rationales behind the choices of primers and reaction conditions were as follows.

1. PCR priming sites should be located in introns rather than exons. By using intronic regions as priming sites greater flexibility is permitted in choosing the size of the region to be amplified, facilitating resolution of reaction products on agarose gels. In addition, this enables detection of sequence alterations flanking exons that disrupt RNA splicing. Amplifying from within exons would necessitate acrylamide gel electrophoresis for resolution of reaction products as dystrophin, like most genes, contains a very narrow range of exon sizes. Either choice necessitated the isolation and sequencing of regions flanking deletion-prone exons of the gene, for although the complete human dystrophin cDNA sequence has been published, less than a third of the splice sites were known and flanking sequences had been reported for only two exons. To date, we have sequenced approximately 1 kb of DNA flanking each of 11 separate DMD exons. Six of these regions have been combined with a seventh (kindly provided by S. Malhotra,

R. Worton, and P. Ray) and with the two previously published intronic regions to develop the current nine-plex reaction[25] (Chamberlain, manuscript in preparation).

2. We have routinely used oligonucleotides 23-28 bases long for multiplex amplification. Although we have not rigorously tested mixtures of varying length primers, we reasoned that longer oligonucleotides would permit higher stringency annealing, thus providing greater specificity during multiplex amplification reactions. The G/C content of primers generally has varied from 40-50%. Lower G/C contents have led to poor amplification while higher contents have occasionally led to spurious amplification products. This is conveniently within the G/C content range generally found in mammalian introns.

3. The reaction buffer system that has permitted multiplex amplification contains 10% (V/V) dimethylsulfoxide (DMSO).[13] Although the presence of organic solvents may not be optimal for most uses of *Taq* polymerase, we have not achieved efficient multiplex amplification in an aqueous buffer. Additional parameters that have been varied include the amount of enzyme used, the amount of template DNA, the concentrations of Mg^{++}, dNTPs and primers, and the time and temperature of annealing and DNA polymerization steps. The current protocols are listed in Protocol D. However, it should be emphasized that we are continually modifying these conditions, in particular after the addition of each new primer set and that not all parameters have been rigorously varied in relation to all others. Although the current concentrations of Mg^{++}, dNTPs and primers appear optimal, as long as their ratios remain constant, the absolute amounts of these reagents can be varied without dramatic effects on the reactions. Different combinations of primers have been observed to require different conditions for optimal multiplex amplification and it may be necessary to optimize reaction conditions for any given set of oligonucleotides. Finally, as more primer sets are added, the permissive reaction conditions become increasingly less flexible.

4. The reaction conditions have been designed to minimize the possibility of false positive amplification. The use of PCR for deletion detection could lead to serious errors if amplification from non-targeted regions of the genome or contaminating exogenous or maternal DNA were to occur. Each primer set is tested individually and with all of the others to ensure that amplification occurs only from the specifically targeted genomic sequences. Although two primers may be specific when used alone, spurious amplification can result when 16 additional primers are added. For DMD, the specificity of reactions is easily checked by using template DNA containing partial dystrophin gene deletions that have been previously delineated *via* Southern analysis with full-length cDNA clones. Of greater concern is the possibility of reagent or sample contamination with exogenous or maternal DNA. Reagent contamination is minimized in two ways. First, the preparation of large quantities of individually aliquoted reaction mixes ("kits") allows

large batches of reagents to be checked for contamination prior to use. Second, we physically separate the preparation and analysis stages of the reactions. Thus, amplified reactions are opened and aliquots removed for analysis at a separate location from where the reactions are initiated. In addition, separate pipettors are used to sample amplified reactions than are used to mix together the initial ingredients. These precautions are critical to prevent minute quantities of prior reaction products from serving as efficient template for future reactions.

Sample contamination must also be kept to a minimum. Equipment that has been in contact with either prior reaction products or cloned DNA complementary to any of the regions targeted for amplification must not be used to prepare template DNA. We prepare all DNA samples for PCR analysis on an Applied Biosystems model 340A DNA extractor. Preparation of several hundred samples on this machine has not yet led to any detectable contamination from either exogenous or prior sample sources. Maternal DNA contamination of amniotic fluid cells or chorionic villus specimens (CVS) is a problem less easily controlled by laboratory personnel. CVS routinely should be dissected of maternal decidual tissue prior to extraction of DNA. Beyond this, care must be taken to ensure that amplification is kept to the absolute minimum required to make a diagnosis. By performing reconstitution experiments with normal and partially deleted DNA samples, we have observed that levels of maternal DNA at up to 5% of the total will not lead to false positive amplification as long as the reactions do not approach saturation[27] (Figure 6B). We have not yet observed false-positive amplification during analysis of approximately 25 amniotic fluid cells or CVS DNA samples.

DETECTION OF CARRIERS OF NEW MUTATIONS

The direct detection of mutant alleles in LN and DMD males is relatively simple compared to the assessment of female carrier status which is complicated by the presence of normal alleles. Broadly, there are two strategies that may allow carrier detection. The first involves the direct detection of the recessive allele in the heterozygote, but the precise approach may differ depending on whether the mutation has already been characterized in the hemizygous affected male. The second method of carrier detection is DNA linkage analysis. The PCR-based techniques that may be adapted for linkage studies are the same as the methods used for direct mutation detection, except that the DNA base differences that are assayed are polymorphic markers in or near the mutant gene rather than the disease mutation itself.

Multiplex DNA amplification and automated direct DNA sequencing each have the potential for direct detection of new mutant alleles in heterozygotes. The multiplex amplification method relies on quantitative representation of the starting material in the PCR throughout the reaction cycles. In females who are heterozygous for deletions or duplications within the DMD gene, the effective template DNA concentration will vary by two-fold in the affected regions. By ensuring that amplification reactions do not approach saturation, we have occasionally observed half or double the relative yield of individual amplified fragments. Examination of these female DMD carriers *via* quantitative Southern analysis has confirmed the presence of heterozygous DNA duplications or deletions, respectively (Fenwick *et al.*, manuscript in preparation). We are currently optimizing the conditions necessary for reproducible discrimination of gene dosage to permit routine screening of females for heterozygous gene rearrangements.

The detection of mixtures of alleles that differ by a single DNA base by direct DNA sequencing requires the signal from two bases at a single site to be reliably distinguished from noise that may occur from non-specific termination events in the dideoxynucleotide sequencing reactions. Examples of heterozygous detection by manual direct DNA sequencing have been reported at loci where the possible DNA alterations were known before the assay.[21,28,29] To our knowledge, there are no reports of heterozygous new mutations identified for the first time by the method, nor of the application of direct automated DNA sequencing to the problem. Reconstruction experiments are underway to test the feasibility and reliability of this approach.

A single DNA base change that is identified in an affected male with an X-linked recessive disease can be detected in other family members by a variety of PCR based methods. Occasionally, the base change will alter a restriction endonuclease recognition sequence and normal and mutant alleles can be distinguished by restriction enzyme digestion and agarose gel electrophoresis of an amplified fragment containing the mutation site.[13] More frequently, no restriction endonuclease recognition site is involved and mutation detection requires the synthesis of oligonucleotides that are specific to each of the alleles.[30-33] The allele specific oligonucleotides (ASOs) can then be labeled either with a radioisotope or a non-radioactive reporter and sequentially hybridized to PCR-amplified fragments from each of the family members.[31] Stringent washing of the hybrids distinguishes the binding of perfectly matched oligomers from hybrids that contain a mismatch because of the mutation. Identification of oligonucleotides that bind at the most stringent conditions implies the mutation sequence.

As an alternative to ASO probing, we have developed a procedure for detection of known single DNA base changes based on competition between two oligonucleotide primers for DNA synthesis.[33] Competitive oligonucleotide priming (COP) is based on the strong preference of a DNA template for a perfectly matched oligonucleotide primer, relative to a primer with a single DNA base difference. Oligonucleotide primers are synthesized to provide perfect matches with either the normal or the mutant alleles. The two primers are mixed together with the DNA template and the preferentially bound oligonucleotide is extended from the 3' terminus by a DNA polymerase. The signal from this reaction can be amplified *via* PCR using a third "common" primer to simplify detection of the differentially labeled COP primers. Under appropriate conditions, a greater than 100-fold difference in signal intensity results from the radioactive incorporation of a correctly matched DNA primer than a mismatched primer.

Effective COP primers generally are shorter than oligonucleotides that are commonly used for direct PCR amplification of unique mammalian gene sequences.[33] Therefore, it is necessary to "preamplify" the region of interest before performing the COP assay. Generally, one of the initial PCR primers can function as the common primer in the COP assay and so a total of four oligonucleotides must be synthesized for each allelic pair. In order to assay one sample for two possible alleles, two separate reactions with radioactive oligonucleotides must be performed. The reaction products are analyzed by agarose gel electrophoresis and ethidium bromide staining to ensure the correct size COP product has been amplified. The gel is then exposed to X-ray film to determine which of the reactions has incorporated the radiolabeled primer. Identification of the incorporated primer implies the DNA template sequence and the genotype of the individual from whom the sample was derived.

Detection of single DNA base differences in PCR-amplified human DNA fragments by restriction endonuclease digestion, ASO hybridization, and COP are shown in Figure 7. The COP procedure is, in our opinion, operationally simpler than ASO hybridization and more generally applicable than restriction endonuclease digestion. As COP can be performed without the need to bind DNA to filter supports, the prospects for automation are greatly enhanced. Consequently, COP may become the preferred DNA sequence detection method for widespread genetic screening programs.[34,35]

GENERAL STRATEGIES FOR MUTATION DETECTION

Molecular diagnosis for all HPRT and DMD affected families will be greatly simplified when the individual methods developed at each locus are integrated. For example, the current practice of screening new LN cases for DNA deletions by Southern analysis will be obviated when sufficient HPRT gene sequence is available to allow construction of multiplex primer sets that amplify each HPRT exon. The PCR amplification of mRNA can facilitate the identification of DMD point mutations, although the large size of the dystrophin message means that heteroduplex mapping of the amplified product may be necessary before direct sequencing of the mutated region. Therefore, point mutation detection methods are being focussed on the identification of additional polymorphic DNA markers so that PCR-based high resolution haplotype analysis may be more generally used to diagnose DMD family members. Similar searches for polymorphic DNA markers within the HPRT locus are underway.

When new DNA sequence polymorphisms are identified, additional oligonucleotide primers can be incorporated into multiplex reactions so that DNA deletion screening and amplification of informative regions may be performed simultaneously. This greatly simplifies the choice of strategies for the diagnostician, who can use the multiplex assay for a primary screen for DNA deletions and, in the case of a negative result, assay the same reaction products to perform either direct mutation detection or linkage analysis. The final choice of linkage *vs.* direct mutation detection will depend on the availability of family members, the number of currently

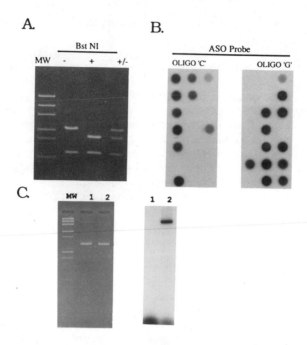

A.

B.

C.

Figure 7. PCR-mediated DNA haplotyping. Shown are three separate methods for the identification of single DNA base differences in human DNA. In each case, a 942-base fragment flanking exon 17 (Figure 5B) was PCR-amplified and analyzed as follows. **A.** Digestion with restriction endonuclease *Bst*NI; 5 µl of the reaction was digested with the enzyme and analyzed by agarose gel electrophoresis; (-) male DNA lacking the polymorphic *Bst*NI recognition sequence, (+) male DNA containing the polymorphic *Bst*NI recognition sequence, (+/-) heterozygous female DNA. **B.** ASO hybridization analysis of a G *vs.* C DNA sequence polymorphism in the same region shown amplified in A. DNA samples from 20 male and 1 heterozygous female were analyzed by dot-blot hybridization using 15-base oligonucleotides specific for the "C" (left) or "G" (right) containing alleles. Only DNA from the female heterozygote hybridizes with both probes. **C.** COP analysis of a single DNA base difference in PCR-amplified human genomic DNA. The polymorphic site, amplified DNA, and the oligonucleotide primers were those used in the ASO analysis shown in (B). 5% of the products from amplification of one male DNA sample was mixed with the two allele-specific COP primers and a third "common" primer that enabled PCR amplification of a 377-base fragment. Two reactions were performed, each with one of the allele-specific primers radiolabeled. The expected size COP product is shown analyzed by agarose gel electrophoresis and ethidium bromide staining at left. When the gel was exposed to X-ray film, the preferential incorporation of the perfectly matched radiolabeled oligonucleotide primer was revealed (right).

known polymorphic markers in the region, and the size of the mutation target.

The application of the techniques described here for the analysis of LN and DMD can be readily adapted for the diagnosis of other genetic diseases that frequently arise from new mutations and for which the DNA sequence of critical regions in the normal genes are known. The assays can also be readily combined with tests for frequently inherited disorders that result from common alleles and, thus, the opportunity for comprehensive genetic screening by DNA based methods now appears within reach.

PROTOCOLS

A. cDNA Synthesis

1. Mix on ice; 1-5 µg total cellular RNA, 0.5 µl (14U) RNasin (Pharmacia), 2.0 µl (10 µg) pd(N)6 primers (Pharmacia), 4 µl of 5 X pol buffer (250 mM Tris.HCl, pH 8.3 at 37°C, 40 mM MgCl$_2$, 150 mM KCl, 50 mM DTT), and H$_2$O (DEPC treated) to 15.5 µl.

2. Heat at 95°C for 1 min, chill on ice, pulse/spin in microfuge then add at room temperature: 2.0 µl of dNTP's (mix of 25 mM each of dATP, dTTP, dCTP, dGTP), 0.5 µl RNasin, 1.0 µl (12 U) M-MuLV Reverse Transcriptase (Pharmacia).

3. Incubate 37°C for 1 hr, then add 30 µl of 0.7M NaOH, 40 mM EDTA, mix gently, and incubate at 65°C for 10 min.

4. Add 5 µl of 2 M ammonium acetate, pH 4.5, mix, add 130 µl of ethanol, chill, spin, wash in 70% ethanol, 100% ethanol, dry.

B. PCR Amplification of cDNA and Production of Single Strands

1. Mix 5-10% of the product of one cDNA synthesis reaction with 100 pmols of each PCR primer in a total volume of 100 µl containing 6.7 µM MgCl2, 16.6 mM (NH$_4$)$_2$SO$_4$, 5 mM β-mercaptoethanol, 6.8 mM EDTA, 67 mM Tris.HCl, pH 8.8 at 25°C, 80 mg/ml BSA, 1 mM of each dNTP (4 µl of 25 mM mix) and 10% DMSO.

2. Heat to 95°C for 7 min and briefly centrifuge.

3. Add 5 units of *Taq* DNA polymerase (Perkin-Elmer/Cetus).

4. We typically perform 30 cycles of DNA polymerization (65°C, 1-3 min) denaturation (92°C, 30 sec) and annealing (37-55°C, 30 sec). The final 65°C incubation is extended for 7 min.

5. Analyze double-stranded PCR product by agarose gel electrophoresis.

6. Take 1 μl of the PCR product to initiate single strand producing reactions, using only one primer. The primer that is used is in opposite sense to the sequencing primer that will be employed. Apart from the absence of one primer, the single strand producing reactions are identical to the usual PCR.

7. Do 30 cycles of the single strand-producing reactions.

8. Dilute the reaction with an equal volume of H_2O and add the same volume of 7.5 M ammonium acetate, mix, add 2.5 vols of ethanol, chill 15 min at -70°C, and spin for 15 min in a microfuge. Repeat the ammonium acetate precipitation. Wash with 70% ethanol, 100% ethanol and dry to completion under vacuum. Dissolve in 10 μl immediately before use in the DNA sequencing reaction.

9. Occasionally, step (8) results in coprecipitation of salt and protein from the reaction mix. If this is troublesome, the single strand-producing reactions may be carried out in the buffer recommended by Perkin-Elmer/Cetus; however, the optimum concentration of primer, dNTPs, and Mg^{++} may need to be determined.

C. DNA Sequencing Reactions

I. Manual Sequencing

1. Radiolabel oligonucleotide primer by mixing 100 pmols of primer, 50-70 mCi of [^{32}P] γ-ATP (6000 Ci/mmol) and 30 units of T4 polynucleotide kinase in 10 mM Tris.HCl, pH 7.6, 10 mM $MgCl_2$, 1 mM DTT, in a total volume of 50 μl. Incubate at 37°C for 40 min and purify the labeled primer by passage through an NENsorb column (NEN DuPont). Dry the product to completion under vacuum, then resuspend in 12 μl of H_2O immediately before use.

2. Add 5.0 μl of single strand DNA template to 3 μl of labeled primer and 2.0 μl of 5 X pol buffer (see Protocol A) in a standard 1.5-ml microcentrifuge tube.

3. Heat to 95°C for 10 min and pulse-spin in a microfuge to bring down condensation.

4. Dispense 2.5 μl aliquots of the primer template mixture into four appropriately labeled tubes (1T, 1C, 1G, etc.). Do this step on the bench.

5. Add 2.0 μl of the appropriate dideoxyterminator/Sequenase mix (80:8 μM [USB] with 0.5 U/μl Sequenase [USB] added just before use) to each of the four tubes and place immediately at 50°C. Incubate for 40 min.

6. Pulse-spin in a microfuge to bring down condensation and add 3.0 μl of STOP solution (95% formamide, 20 mM EDTA, 0.05% bromophenol blue, 0.05% xylene cyanol). Heat to 80°C for 3 min and then analyze by standard electrophoresis and autoradiography.

II. Automated DNA Sequencing

1. Construction of the four dye labeled primers (TAMRA-, FAM-, ROX-, JOE-) is described elsewhere.[20,36] The primers are maintained as 1 mM stocks at -20°C in single use aliquots.

2. Pool products from two single strand-producing reactions in a total volume of 24 μl of H_2O and construct annealing reactions as follows: "A" reaction; 3 μl template, 1 μl JOE primer, 1 μl 5 X Pol buffer (see above): "C" reaction; 3 μl template, 1 μl FAM primer, 1 μl 5 X Pol buffer: "G" reaction; 9 μl template, 3 μl TAMRA primer, 3 μl 5 X Pol buffer: "T" reaction; 9 μl template, 3 μl ROX primer, 3 μl 5 X Pol buffer.

3. Heat mixes to 85°C for 2 min and cool to room temperature over 30 min.

4. Pulse spin and add dideoxy termination/Sequenase mixes (see above) to each tube ("A": 4 μl,"C": 4 μl,"G": 12 μl,"T": 12 μl). Incubate reactions at 50°C for 5 min, heat to 68°C for 10 min, pool and precipitate by addition of 6 μl 3 M NaOAc (pH 5.2) and 170 μl ethanol.

5. Wash in 80% ethanol, dry briefly under vacuum, and redissolve in 12 μl of formamide. Electrophorese 3-5 μl of each sample on an 8% polyacrylamide gel at a constant 27W and 50°C.

D. Multiplex DNA Amplification

1. Prepare template DNA from lymphoblasts, amniotic fluid cells, or CVS dissected of decidual tissue using standard protocols.[10]

2. Add to a 0.5 μl microfuge tube: H_2O to 45 μl final volume, 10 μl 5X *Taq* polymerase buffer (83 mM $[NH_4]_2SO_4$; 335 mM Tris.HCl, pH 8.8; 33.5 mM $MgCl_2$; 50 mM β-mercaptoethanol; 850 mg/ml bovine serum albumin; and 34 μM EDTA), 3 μl 25 mM each dNTP (Pharmacia), 25 pmols of each oligonucleotide primer, and then 5 μl of DMSO. The reaction mixes are stable at -70°C for up to 3 months.

3. Mix gently; add 250 ng template DNA. (Dilute DNA to a final concentration between 50 and 250 ng/μl so that the DNA may be added to the reaction in a volume of 5 μl or less.) Add H_2O to a final volume of 50 μl.

4. Add 5 units *Taq* polymerase (Perkin-Elmer/Cetus), mix gently.

5. Add 25 μl paraffin oil, centrifuge 5 sec.

6. Place sample in an automatic thermocycler. Cycle as follows: A. 94°C X 6 min (once); B. 94°C X 30 sec; 56°C X 30 sec; 65°C X 4 min (repeat 23 times); C. 65°C X 7 min (once); D. 4°C until analysis (up to two months).

7. Electrophorese 15 μl of the reaction products on a 1.4% agarose or a 3% NuSieve agarose gel (FMC Bioproducts) (containing 0.5 μg/ml ethidium bromide) for 2 hr at 3.7 V/cm in 90 mM Tris-base, 90 mM boric acid, 1 mM EDTA.

8. Photograph the gel or otherwise record the results.

ACKNOWLEDGMENTS

We gratefully acknowledge the expert technical assistance of Andrew Civitello, Nancy Farwell, Donna Muzny, Phi-Nga Nguyen, Joel Ranier, and collaborators Lincoln McBride and Sandy Koepf of Applied Biosystems, and

John Belmont, Grant MacGregor and Jan Witkowski for many useful discussions. Thanks to John Moyer for the black and white reproduction of Figure 4. C.T.C. is an investigator of the Howard Hughes Medical Institute and R.A.G. is a recipient of the Muscular Dystrophy Association's Robert G. Sampson Distinguished Research Fellowship. This work is based in part upon work supported by the Texas Advanced Technology Program under Grant No. 3304. We also gratefully acknowledge the support of the Muscular Dystrophy Association's Task Force on Genetics and Public Health Service grant #DK31428.

REFERENCES

1. Kelley, W.N., and Wyngaarden, J.B. (1983) in *Metabolic Basis of Inherited Disease*, Stanbury, J.B., Wyngaarden, J.B., Fredrickson, D.S., Goldstein, J.L., and Brown, M.S., eds. McGraw-Hill, 5th ed., pp. 1115-1143.
2. Emery, A.E.H. (1987) *Oxford Monographs on Medical Genetics*, No 15. Oxford University Press.
3. Patel, P.I., Framson, P.E., Caskey, C.T., and Chinault, A.C. (1986) *Mol. Cell. Biol.* 6:393.
4. Chamberlain, J.S. *et al.* (1988) *Science* 239:1416-1418.
5. Hoffman, E.P. (1988) *Neuron* 1:411-420.
6. Kelly W.N. *et al.* (1969) *Ann. Intern. Med.* 70:155-206.
7. Koenig, M. *et al.* (1987) *Cell* 50:509-517.
8. Yang, T.P. *et al.* (1984) *Nature* 310:412-414.
9. Wilson, J.M. *et al.* (1986) *J. Clin. Invest.* 77:188-195.
10. Ward, P.A. *et al.* (1989) *Am. J. Hum. Genet.* 44:270-281.
11. Mullis, K., and Faloona, F.A. (1987) *Meth. Enzymol.* 155:35-40.
12. Saiki, R.K. *et al.* (1988) *Science* 239:487-491.
13. Kogan, S.C., Doherty, M., and Gitschier, J. (1987) *N. Engl. J. Med.* 317:985-990.
14. Saiki, R. *et al.* (1985) *Science* 230:1350-1354.
15. Gibbs, R.A., and Caskey, C.T. (1987) *Science* 236:303-305.
16. Veres, G., Gibbs, R.A., Scherer, S.E., and Caskey, C.T. (1987) *Science* 237:415-417.
17. Myers, R.M., and Maniatis, T. (1986) *Cold Spring Harbor Symposia on Quantitative Biology* 51:275-84.
18. Chirgwin, J.M., Przybyla, A.E., McDonald, R.J., and Rutter, W.J. (1979) *Biochemistry* 18:5294-5299.
19. Kawasaki, E., this volume.
20. Gibbs, R.A., Nguyen, P.N., McBride, L.J., Keopf, S.M., and Caskey, C.T. (1989) *Proc. Natl. Acad. Sci. USA* 86:1919-1923.
21. Gyllensten, U., and Erlich, H.A. (1988) *Proc. Natl. Acad. Sci. USA* 85:7652-7656.
22. Innis, M.A., Myambo, K.B., Gelfand, D.H., and Brow, M.A. (1989) *Proc. Natl. Acad. Sci. USA* 85:9436-9440.
23. Smith, L.M. *et al.* (1986) *Nature* 321:674-679.
24. Connell, C. *et al.* (1987) *Biotechniques* 5:342-348.
25. Edwards, A.O. *et al.* (1988) *Am. J. Hum. Genet.* 43:A182.
26. Chamberlain, J.S. *et al.* (1988) in *Cellular and Molecular Biology of Muscle Development*, Stockdale, F. and Kedes, L., eds. UCLA Symposia on Cellular and Molecular Biology, New Series, Vol. 93:951-962.

27. Chamberlain, J.S., Gibbs, R.A., Ranier, J.E., Nguyen, P.N., and Caskey, C.T. (1988) *Nucl. Acids Res.* 16:11141-11156.
28. Newton, C.R. *et al.* (1988) *Nucl. Acids Res.* 16:8233-8244.
29. Wong, C., Dowling, C.E., Saiki, R.K., Higuchi, R.G., Erlich, H.A., and Kazazian, Jr., H.H. (1987) *Nature* 330:384.
30. Kidd, V.J., Wallace, R.B., Itakura, I., and Woo, S.L.C. (1983) *Nature* 304:230-234.
31. Saiki, R.K., Bugawan, T.L., Horn, G.T., Mullis, K.B., and Erlich, H.A. (1986) *Nature* 324:163-166.
32. Landegren, U., Kaiser, R., Sanders, L., and Hood, L. (1988) *Science* 241:1077-1080.
33. Gibbs, R.A., Nguyen, P.N., and Caskey, C.T. (1989) *Nucl. Acids Res.* In press.
34. Caskey, C.T. (1987) *Science* 36:1223-1229.
35. Gibbs, R.A., and Caskey, C.T. (1989) *Ann. Rev. Pub. Health* 10:27-48.
36. Smith, L.M., Fung, S., Hunkapiller, M.W., and Hood, L.E. (1985) *Nucl. Acids Res.* 13:2399-2412.
37. Baumbach, L.L., Chamberlain, J.S., Ward, P.A., Farwell, N.J., and Caskey, C.T. (1989) *Neurology*, in press.

CHAPTER 16

HLA Class II Gene Polymorphism: DNA Typing, Evolution, and Relationship to Disease Susceptibility

Henry A. Erlich and Teodorica L. Bugawan

INTRODUCTION

The detection of HLA class II polymorphism is valuable in the areas of individual identification, tissue typing for transplantation, and genetic susceptibility to specific autoimmune diseases. Polymorphism in the HLA class II region (see Figure 1 for map) has been identified using serologic reagents (HLA-DR and -DQ specificities), by cellular techniques (Dw and DPw specificities) and, more recently, by restriction fragment length polymorphism (RFLP) analysis. For HLA class II typing, RFLP analysis is based on the presence or absence of polymorphic restriction sites located primarily in non-coding regions which are in linkage disequilibrium with allelic variation in coding sequences. Until recently, the direct analysis of coding sequence polymorphism has been difficult. However, the enzymatic amplification of specific DNA sequences using the PCR has provided a new approach to genetic typing.[1-4] The capacity of the PCR to amplify a specific segment of genomic DNA has made it an invaluable tool in the

HLA Class II Genes

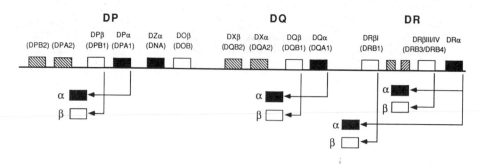

HLA Class II Proteins

Figure 1. Map of HLA class II region. The α-chain loci are represented as filled-in boxes and the β-chain loci as open boxes. The unexpressed loci are represented as hatched boxes. In the new nomenclature for the HLA class II loci,[19] the DQα locus is designated DQA1 and the DXα is designated DQA2. To avoid confusion with the allelic nomenclature reported in ref. 12, we have retained the previous locus designations in the text. In this figure, the new class II locus designations are shown in parentheses.

study of polymorphism and evolution, as well as in the analysis of genetic susceptibility to disease. In all of these areas, a particular gene must be examined in a variety of individuals; either within a species, in different closely related species, or in patient and in healthy control populations. We have used PCR, initially with the Klenow fragment of *E. coli* DNA polymerase I and more recently with the thermostable *Taq* DNA polymerase, to determine the allelic sequence diversity of the HLA class II genes (HLA-DRβ, HLA-DQα, HLA-DQβ, and HLA-DPβ). Since serologically defined alleles at some of these loci have been associated with specific autoimmune diseases (e.g., insulin dependent diabetes), we have also compared the distribution of allelic DNA sequences at these class II loci in patients and in controls. Once the allelic sequences were determined, their frequencies in the patient and control populations were measured using oligonucleotide probes to analyze PCR amplified DNA.

We have also examined the evolution of the class II polymorphism in the contemporary human population by comparing the sequences of PCR amplified class II gene segments from several different individuals from a variety of primate species. Here, we discuss briefly the evolution of class II polymorphism, its relation to disease susceptibility, and present a new and powerful method for the rapid genotyping of amplified DNA samples.

CLASS II SEQUENCE POLYMORPHISM

The HLA-D or class II genes are organized into three distinct regions, HLA-DR, -DQ, and -DP, each of which encode an alpha and beta glycopeptide. The association of these alpha and beta chains forms a heterodimeric transmembrane protein expressed on a number of cell types, including B lymphocytes, macrophages, and activated T lymphocytes. These highly polymorphic proteins bind peptide fragments of foreign antigen and it is the resulting complex which is recognized by the T cell receptor, leading to activation of the T lymphocyte. The polymorphism of class II genes is localized to the NH_2 terminal outer domain encoded by the second exon. Using PCR primers to conserved regions, we have amplified and sequenced the second exon of these class II loci from many different individuals, revealing a remarkable degree of allelic diversity (Figures 2-4). (The sequence polymorphism in the DRβ loci and the oligonucleotide probes for DR-typing will be reported elsewhere.) These class II sequences were determined either by M13 cloning of the amplified product followed by chain termination sequencing of the purified single-stranded phage DNA or by direct sequencing using the asymmetric primer method[5] to generate single strands from the PCR. This approach has shown that, in general, the serologically defined allelic variants are quite heterogeneous at the sequence level. For example, the DRw6, DQw1 haplotype can contain any one of three different DRβI, three different DRβIII, six different DQβ, and three different DQα DNA sequence-defined alleles.[6]

A variety of analytic techniques has been used to detect genetic variation in PCR-amplified DNA, including restriction enzyme digestion, RNase A cleavage, denaturing gradient gel electrophoresis, sequence analysis, as well as allele-specific oligonucleotide (ASO) primers and ASO hybridization probes. For the analysis of amplified loci with many allelic variants, we have found that the use of non-radioactively-labeled ASO or sequence-specific oligonucleotide (SSO)[††] probes is the most general and convenient approach. The use of these probes in a dot-blot procedure is a powerful and rapid genetic testing method that has been used for the diagnosis of sickle-cell anemia[4] and β-thalassemia[7] as well as for HLA-DQα genotyping for the identification of forensic samples.[8] This approach involves the PCR amplification of a specific region (i.e., the second exon of the DQβ locus) and the subsequent immobilization of the amplified DNA to replicate filters (nylon membranes). Each filter is then hybridized with an SSO probe labeled either with ^{32}P, biotin, or with an enzyme such as horseradish peroxidase.[8] If labeled with an enzyme, the bound probe is detected by the

---

[††]In some cases, the hybridization of an oligonucleotide probe does not uniquely specify an allele because the specific sequence is present in more than one allelic variant. In the absence of allele-specific sequences, a given allele is identified as a pattern of SSO probe binding (see section on DPβ typing).

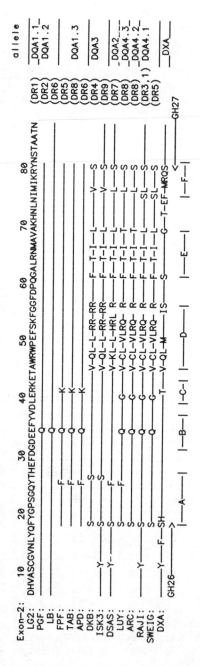

Figure 2. HLA-DQα protein sequences. The DQα allele designations are shown on the right. The names of representative cells are shown to the left and their DR serotypes to the right of the sequences. Polymorphic segments corresponding to specific oligonucleotide probes are designated A through F.

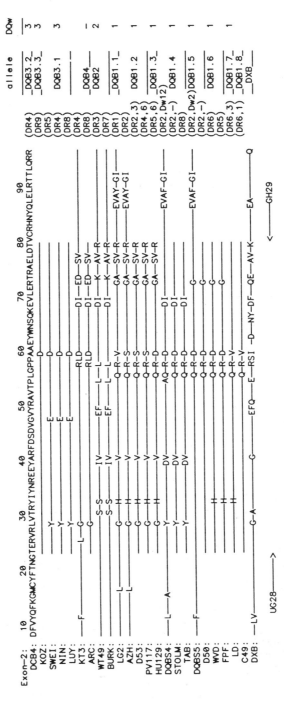

Figure 3. HLA-DQβ protein sequences. The DQβ allele designations and the corresponding DQw specificities are shown on the right. The names of representative cells are shown to the left and their DR serotypes to the right of the sequences. DQB3.1 corresponds to the newly defined serologic specificity DQw7, DQB3.2 to DQw8, and DQB3.3 to DQw9. The DQB1.1, 1.2, and 1.3 alleles correspond to DQw5 and the DQB1.4, 1.5, 1.6, 1.7, and 1.8 alleles correspond to DQw6.

```
                   10        20        30        40        50        60        70        80        90      allele   DPW
 Exon-2:  ---------------------------------------------------------------------------------------       ------   ---
    HHK:  NYLFQGRQECYAFNGTQRFLERYIYNREEFARFDSDVGEFRAVTELGRPAAEYWNSQKDILEEKRAVPDRMCRHNYELGGPMTLQRR      DPB4.1    4
    APD:    ----------------------------V------------------DE------------------------------------      DPB4.2    4
     LB:  ------------------------------V------------------DE------------E-----------------------      DPB2.1    2
    QBL:    ---------------------------LV------------------E-------------E-----------------------      DPB2.2    2
   PIAZ:    ----------------------------V------------------DE------------E------V-------DEAV-----      DPB8      -
    JRA:    ----------------------------V------------------DE------------E--------------DEAV-----      DPB16     -
    HAS:    ---------------------------LV------------------E----------------------------DEAV-----      DPB5      5
   CB6B:    ----------------------------V------------------E-------------E------I-------DEAV-----      DPB19     -
   RAJI:  ----------------------------------------------------------------------V-------DEAV-----      DPB7      -
    SLE:  --VY-L------------------------V------------------DED-------L----------V-------DEAV-----      DPB3      3
   JMOS:    VY-L------------------------V------------------DED-------L---E--------------DEAV-----      DPB6      6
    CRK:    VY-L---------------------Q-Y-----------------------------L---R--------------DEAV-----      DPB11     -
     NB:    VY-L-----------------------Y---------------------------------E------I-------DEAV-----      DPB13     -
    LUY:    VY-------------------------Y----------------------------------------V-------DEAV-----      DPB1      1
    JCA:    VY--------------------------V------------------DE---------------------------V--------      DPB18     -
    PLH:    VY-----------------------Q-Y-----------------------------L---R--------------V--------      DPB15     -
   8268:    VH-L------------------------V------------------DED-------L----------V-------DEAV-----      DPB14     -
   BM21:    VH-L------------------------V------------------DE------------E------V-------DEAV-----      DPB10     -
    TOK:    VH-L------------------------V------------------DED-----------E------V-------DEAV-----      DPB9      -
  AKIBA:  --VH-L-----G---------S--------V------------------DED-----------E------V-R-----DEAV-----      DPB12     -
   JRAB:    VH-L------------------------V------------------DED-----------E--------------DEAV-----      DPB17     -
 PRIESS:  -SVY-E------------VVDGL------YVH---A----L--M-----IG--F-----FM-R---EV-KV---K---ME-LIR---      DXB
UG21--->                                                                                      <---UG19
           |-A-|                  |-B-|               |-C-|    |-D-|  |-E-|   |-F-|
```

Figure 4. HLA-DPβ protein sequences. The DPβ allele designations and the corresponding PLT-defined DPw specificities are shown on the right. The names of representative cells are shown to the left, polymorphic segments corresponding to specific oligonucleotide probes are designated A through F.

enzymatic conversion of a colorless soluble substrate to a colored precipitate.

For a locus with n alleles, each amplified sample must be immobilized on n membranes and each membrane hybridized to one of n labeled probes. Thus, the procedural complexity of this approach is a function of the number of oligonucleotide probes required for complete genetic analysis. To address this problem, we have recently developed a "reverse dot-blot" procedure in which the oligonucleotide probe is immobilized on a membrane and hybridized to a labeled PCR product (Figure 5).[9] In this method, a panel of oligonucleotide probes is tailed with polydT using terminal transferase and then UV-cross-linked to a nylon membrane. The PCR product, labeled during amplification by using biotinylated primers, is then hybridized to the immobilized array of oligonucleotide probes. The presence of the specifically bound PCR product is detected using a streptavidin-horseradish peroxidase conjugate. This method, described in detail by Saiki *et al.*[9] has been applied to detection of a variety of β-thalassemia mutations and HLA-DQα alleles. Both the dot-blot and the reverse dot-blot method represent rapid and precise approaches for typing HLA class II polymorphism.

EVOLUTION

The extensive polymorphism observed in the human class II loci could have been generated by recent (i.e., after speciation) mutation, recombination or gene conversion followed by selection for the newly arisen variants. Alternatively, ancient (i.e., before speciation) polymorphism maintained by selection could have given rise to the observed allelic diversity. To test this hypothesis, the polymorphic second exon of the DQα locus was amplified and sequenced from a number of individuals in a variety of primate species: the hominoids (humans, chimpanzees, gorillas,) some old world monkeys (baboons, rhesus, langur), and some new world monkeys (capuchin and marmoset).[10] Phylogenetic analysis of these sequences was carried out using the method of maximum parsimony. In general, a given human DQα allelic type (e.g., DQA4) is more closely related to its gorilla or chimpanzee allelic counterpart than it is to any other human allelic type (e.g., DQA1, DQA2, or DQA3). This pattern in which the sequences cluster by allelic types rather than by species suggests that most of the DQα allelic diversity in the contemporary human population was present in an ancestral species that gave rise to the human, chimpanzee, and gorilla lineages. Thus, at the DQα locus, the polymorphic alleles (DQA1, DQA3, and DQA4) appear to be ancient (5-20 myr old, based on the known divergence times of the hominoids and old world monkeys) and have been maintained by selection (e.g., overdominance or frequency dependent mechanisms). The DQA2 allele appears to have arisen in the human lineage and may have been generated by a recombination between DQA3 and DQA4.[10] At the DQβ locus, the major allelic types (e.g., DQB1, DQB2, DBB3, and DBB4) also appear to be ancient but, here, some subtypic diversification appears to have occurred *after* speciation (Gyllensten *et al.*, unpublished results).

Format II

Immobilized ASO Probes

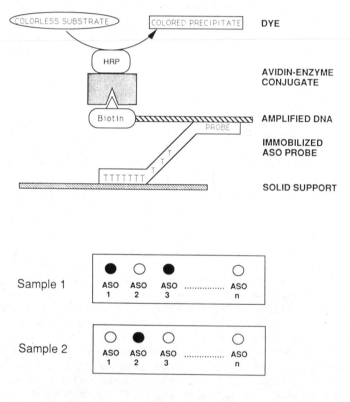

Figure 5. Schematic diagram of immobilized oligonucleotide probe detection of amplified DNA. **A.** An allele-specific probe is "tailed" with a dT homopolymer and immobilized on a solid support. The amplified PCR product, which has incorporated a biotinylated primer, hybridizes to the probe. After washing away unbound DNA and primers, the biotinylated, amplified DNA binds an avidin horseradish peroxidase conjugate. The enzyme then converts a colorless dye into a colored precipitate. **B.** A format for detecting specific alleles in samples of amplified DQα DNA, from heterozygous individuals.

DISEASE SUSCEPTIBILITY

Another area where PCR-based analysis of class II polymorphism has proved very valuable is the study of genetic susceptibility to HLA-associated diseases. A variety of diseases have been associated with specific serologically defined variants (e.g., insulin dependent diabetes mellitus (IDDM) with DR3 and DR4). A disease associated marker is simply one whose frequency is increased in patients relative to controls (e.g., 90% of IDDM patients are DR3 or DR4 compared to 40% of controls). PCR amplification facilitated the sequence analysis of class II alleles derived from patients and controls and the comparison of their distribution in the two populations by oligonucleotide probe hybridization (dot-blot analysis; see above). Since the serotypes (e.g., DRw6) have proved to be genetically heterogeneous, sequence based analysis has revealed specific alleles that are more highly associated with a particular disease. For example, the DRw6 serotype has a relative risk of 2.5 for the autoimmune dermatologic disease, *Pemphigus vulgaris* (PV), whereas a particular DQβ allele, DQB1.3 (one of six DQβ alleles on DRw6, DQw1 haplotypes) has a relative risk of ~100.[11] Unlike serologic or RFLP analysis of class II polymorphism, PCR based analysis reveals not only that two alleles are different but *how* and *where* they differ. In the example above, the susceptible DQB1.3 allele differs from another non-susceptible allele by only a Val to Asp substitution at position 57, implicating this residue as a critical element in susceptibility. (It should be noted that other non-susceptible DQβ alleles also encode Asp at position 57 so that it is the entire *allele* rather than just an isolated residue that confers susceptibility).

By comparing the sequences of susceptible and non-susceptible alleles, the amino acid at position 57 of the DQβ chain and the amino acids around position 70 of the DRβI and DPβ chains have been revealed as critical residues in susceptibility to IDDM,[12,13] PV,[6,11,14] Celiac disease,[15,16] and pauciarticular juvenile rheumatoid arthritis.[17] Interestingly, the evolutionary analysis of DQβ polymorphism in non-human primates (see above) has revealed a balanced polymorphism between Asp-57 and Ala, Val, or Ser at this position in *all* primate class II β chains.[18] The maintenance of this polymorphism suggests, as do the disease susceptibility studies, that this residue is functionally important. A more detailed discussion of the relationship of class II sequence polymorphism, as detected by PCR, and disease susceptibility is presented in Horn *et al.*[12] and Scharf *et al.*[11] In general, specific *combinations* of class II alleles are most highly associated with autoimmune diseases.

DNA TYPING

The amplification of the polymorphic exon of the DQα, DQβ, and DRβ loci can be carried out using the primers and conditions listed in Table 1. We will use the HLA-DQα and the HLA-DPβ loci to illustrate the methodology of HLA typing with oligonucleotide probes. The use of these

Table 1. Sequence of Oligonucleotide PCR Primers for HLA Class II Gene Amplification

HLA Gene	Name	Sequence 5' to 3'	PCR Amplification Conditions		
			Denature	Anneal	Extend
DRβ	GH46 GH50	CCGGATCCTTCGTGTCCCCACAGCACG CTCCCAACCCGTAGTTGTGTCTGCA	96°C	55°C	72°C
DQα	GH26 GH27	GTGCTGCAGGTGTAAACTTGTACCAG CACGGATCCGGTAGCAGCGGTAGAGTTG	96°C	65°C	≥65°C
DQβ	GH28 GH29	CTCGGATCCGCATGTGCTACTTCACCAACG GAGCTGCAGGTAGTTGTGTCTGCACAC	96°C	55°C	72°C
DPα	GH98 GH99	CGCGGATCCTGTGTCAACTTATGCCGC GTGGCTGCAGTGTGGTTGGAACGC	96°C	55°C	72°C
DPβ	UG21 UG19	CGGATCCGGCCCAAAGCCCTCACTC GCTGCAGGAGAGTGGGCGCCTCCGCTCAT	96°C	65°C	≥65°C

The denaturation, annealing, and extension times are 30 seconds and ramping from 96° to 55° to 72°C is programmed in the PECI ThermoCycler for 1 sec. In some cases, (DQα and DPβ) the temperature profile is a 2-step cycle from 65°C to 96°C. Using these conditions, the DQα primers (GH26 and GH27) and the DQβ primers (GH28 and GH29) co-amplify to a limited extent the homologous sequence from the linked but non-expressed loci DXα and DXβ. The DRβ primers GH46 and GH50 amplify *all* DRβ loci not simply the DRβl locus.

genetic markers for the forensic analysis of individual identification is discussed in Chapter 17.

DQα

The eight HLA-DQα alleles (Figure 2) can be classified into four major allelic types (A1, A2, A3, and A4) based on the polymorphism in codons 45 to 56 (region D) and into the subtypes (A1.1, 1.2, 1.3, and 4.1, 4.2, 4.3) based on sequence polymorphism in other regions (B, C, and E). The oligonucleotide probes used to define both the types and subtypes and the hybridization conditions are shown in Table 2. We have used these probes labeled either with ^{32}P, biotin, or horseradish peroxidase in a dot-blot format.[8] As described above, the polymorphism can also be detected in a "reverse dot-blot" or immobilized oligonucleotide probe format (Figure 5). This DQα typing format is described in detail in Saiki *et al.*[9] and will be available commercially from Cetus Corporation.

DPβ

The allelic diversity at the DPβ locus presents more of a challenge for oligonucleotide probe typing because of the dispersed nature of the polymorphic sequences. The second exon contains six variable regions with a limited number of polymorphic residues (n=2-3) at each position. This patchwork pattern of polymorphism suggests that recombination may have played a significant role in generating DPβ allelic diversity. Our approach has been to use a panel of fifteen oligonucleotide probes (listed in Table 3) specific for the sequence variants in four polymorphic regions, with the *pattern* of probe hybridization identifying a specific DPβ allele (see Table 4). There are, in principle, 160 different (4x4x5x2) sequence combinations detectable by this panel of probes. In the population surveyed thus far (n≥300), we have identified 22 different DPβ alleles by oligonucleotide typing and sequence analysis. Some rare alleles (e.g., DPβ9, 12, and 17) cannot be distinguished using this probe panel because these alleles differ in segments outside the polymorphic regions typed by our probes. Another potential ambiguity can result from the same probe reactivity pattern being consistent with two different genotypes (e.g., 2.1/5 or 4.2/19). This ambiguity can be resolved by using the oligonucleotide DB26 (Table 3 and Figure 4) as a "group-specific" amplification primer to test whether a given probe is hybridizing to the "GGPM" group of alleles (DPB2.1, 2.2, 4.1, and 4.2) or the "DEAV" group (all other alleles). This is equivalent to "setting the phase" or determing the "haplotype" of the polymorphic regions defined by the probes.

DRβ

The allelic diversity in the HLA-DR region, not addressed in this chapter, is even greater than the HLA-DQ and -DP polymorphism discussed here. At

Table 2. Oligonucleotide Probes for HLA-DQα Allele Typing

Name	Allele Specificity	ASO Probe Sequence 5' to 3'	Region
RH54	all	CTACGTGGACCTGGAGAGGAAGGAGACTGCCTG	B and C
RH83	1	GAGTTCAGCAAATTTGGAG	D
GH88*	1.1	CGTAGAACTCCTCATCTCC	B
GH76*	all but 1.3	GTCTCCTTCCTCTCCAG	C
GH89	1.2, 1.3, 4	GATGAGCAGTTCTACGTGG	B
GH77	1.3	CTGGAGAAGAGGAGAC	C
RH7	12	TTCCACAGACTTAGATTTGAC	D
GH67	3	TTCCCGAGATTTAGAAGAT	D
GH66	4	TGTTTGCCTGTTCTCAGAC	D
HE46	4.2, 4.3	CATCGCTGTGACAAAACAT	E

* indicates that probe sequence is from the non-coding strand

Dot blots were prepared by immobilizing 5 µl of denatured PCR product onto a nylon membrane (i.e., Biodyne B or Genatran). The DNA were fixed onto the membrane either by baking in a vacuum oven for 1 hr or exposing in a UV light for 5 min. DQα oligonucleotide probes, labeled with HRP as described[7] were added at 0.5 picomole probe per ml. Hybridization solution (5x SSPE, 5x Denhardt's and 0.5% Triton X-100) and incubated with the membranes for 20 min at 55°C. The membranes were washed in 2x SSPE, 0.1% Triton X-100 for 5 min at 55°C. The hybridized membranes were developed colorimetrically to detect the signals as follows: the detection procedure were done at room temperature with moderate shaking. The membranes were incubated for 5 min with Buffer B (137 mM NaCl, 2.7 mM KCl, 1.5 mM KH_2PO_4, 8.0 mM Na_2HPO_4, pH 7.4, 5% (v/v) Triton X-100 1 M urea, and 1% dextran sulfate) followed by 5-min wash with Buffer C (100 mM sodium citrate pH 5). The membranes were incubated again for 10 min in Buffer C with 0.1 mg/ml TMB (3,3',5,5'-tetramethylbenzidine in 100% ethanol) under light exclusion. Hydrogen peroxide is added to a final concentration of 0.0015%. Typically, the blue positive signal appears in 1-5 min. The reaction was stopped by washing the membrane for 5 min in water; if filter background is high, additional washing may be necessary.

Table 3. Oligonucleotide Probes for HLA-DPβ Allele Typing

Name	Sequence Variant	SSO Probe Sequence 5' to 3'	Region	Hybridization/Wash Conditions
DB10	LFQG	GAATTACCTTTTCCAGGGA	A	5X@50/42 water bath
DB11	VYQL	ATTACGTGTACCAGTTACG	A	3X@42/42 air incubator
DB12*	VYQG	CGTCCCTGGTACACGTAAT	A	5X@50/42 water bath
DB22	VHQL	ATTACGTGCACCAGTTACG	A	3X@50/42 water bath
DB13	AAE	CCTGCTGCGGAGTACTG	C	3X@50/42 water bath
DB14*	DEE	CAGTACTCCTCATCAGG	C	5X@42/42 water bath
DB59	EAE	CCTGAGGCGGAGTACTG	C	5X@50/42 water bath
DB17	DED	CCTGATGAGGACTACTG	C	5X@50/42 air incubator
DB18	ILEEK	GACATCCTGGAGGAGAAGC	D	3X@55/42 water bath
DB19	ILEEE	GCTCCTCCTCCAGGATGTC	D	5X@55/42 water bath
DB20	LLEEK	GACCTCCTGGAGGAGAAGC	D	5X@55/42 water bath
DB62	LLEEE	GACCTCCTGGAGGAGGAG	D	3X@50/42 water bath
DB63	LLEER	GACCTCCTGGAGGAGAGG	D	3X@50/42 water bath
DB25	GGPM	CTGCAGGGTCATGGCCCCCG	F	3X@50/42 water bath
DB26	DEAV	CTGCAGGGTCACGGCCTCGTC	F	3X@50/42 water bath

* indicates that sequence is from the non-coding strand

The hybridization solution is 5x or 3x SSPE (as indicated above), 5x Denhardt's and 1% Triton X-100. The membranes containing immobilized PCR product are allowed to hybridize with HRP-labeled sequence specific probe for 45 min to 1 hr at 42°, 50°, or 55°C. The membranes were washed for 10 min at 42°C air incubator or water bath with 0.1x SSPE, 0.2% Triton X-100. A much simpler hybridization procedure is currently being explored. The use of TMACl (tetramethyl ammonium chloride) allows the hybridization to be done at one temperature (55°) for all 15 probes. The hybridization solution contains 3 M TMACl, 0.5% SDS, 10 mM Tris-HCl pH 7.5, and 0.1 mM EDTA. The wash solution is 3 M TMACl, 50 mM Tris-HCl pH 8, and 2 mM EDTA, 20 min at 37°C followed by 10-min wash at 55°C.

Table 4. Determination of DPβ Alleles by Oligonucleotide Probe Hybridization

DPβ allele	A				C				D			F			
	LFQG	VYQL	VYQG	VHQL	AAE	DEE	EAE	DED	I-K	I-E	L-K	L-E	L-R	GGPM or VGPM	DEAV
	10	11	12	22	13	14	59	17	18	19	20	62	63	25	26
1	−	−	+	−	+	−	−	−	+	−	−	−	−	−	+
2.1	+	−	−	−	−	+	−	−	−	+	−	−	−	+	−
2.2	+	+	−	−	−	−	+	−	−	+	−	−	−	+	−
3	−	−	−	−	−	−	−	+	+	−	−	+	−	−	+
4.1	+	−	−	−	+	−	−	−	+	−	−	−	−	+	−
4.2	+	−	−	−	−	+	−	−	+	−	−	−	−	+	−
5	−	+	−	−	−	−	+	−	+	−	−	−	−	−	+
6	+	−	−	−	−	−	−	+	−	−	−	+	−	−	+
7	+	−	−	+	+	−	−	−	+	−	−	−	−	−	+
8	−	−	−	+	−	+	−	−	−	+	−	−	−	−	+
9	−	+	−	−	−	−	−	+	−	+	−	−	−	−	+
10	−	−	−	+	+	+	−	−	−	+	−	−	−	−	+
11	−	+	−	−	−	−	−	−	−	−	+	−	−	−	+
12	−	−	−	+	+	−	−	−	−	+	−	−	+	−	+
13	−	−	−	−	−	−	−	+	−	+	−	−	−	−	+
14	−	−	−	−	+	−	−	−	−	−	−	−	−	−	+
15	−	−	−	+	−	−	−	−	−	−	−	−	+	+	−
16	+	−	+	−	−	+	−	+	−	+	−	−	−	−	+
17	−	+	−	−	−	−	−	−	−	+	−	−	−	−	+
18	−	−	+	−	−	+	−	−	+	−	−	−	−	+	−
19	+	−	−	−	−	−	+	−	−	+	−	−	−	−	+

the DRβI locus, 25 different allelic sequences have been identified thus far. The number of expressed DRβ chain loci is also polymorphic with some haplotypes (i.e., DR1 and DR8) encoding only *one* DRβ chain and others containing *two* expressed loci. This second DRβ chain locus is also polymorphic but less so than the DRβI locus.

CONCLUSION

The extensive allelic diversity at the HLA class II loci revealed by molecular analyses is not readily detected by current immunologic approaches. Unlike immunologic typing, DNA methods are capable of typing cells in which the genes are not expressed and, using PCR, of typing minute amounts of sample. In general, the PCR/ASO method for detecting class II gene polymorphism represents a very precise, rapid, and simple method of HLA typing that should prove valuable for tissue transplantation, individual identification, and for studies of disease susceptibility.

ACKNOWLEDGMENTS

We thank our colleagues in the Human Genetics department, Randall Saiki, Stephen Scharf, Ann Begovich, Russell Higuchi, Sean Walsh, and Robert Griffith for their contributions in developing and applying the HLA typing system described here. We are also grateful to Chu-An Chang for the synthesis of the HRP-labeled oligonucleotides.

REFERENCES

1. Mullis, K.B., and Faloona, F. (1987) *Meth. Enzymol.* 155:335.
2. Saiki, R., Scharf, S., Faloona, F., Mullis, K., Horn, G., Erlich, H.A., and Arnheim, N. (1985) *Science* 230:1350.
3. Saiki, R.K., Gelfand, D.H., Stoffel, S., Scharf, S., Higuchi, R.H., Horn, G.T., Mullis, K.B., and Erlich, H.A. (1988) *Science* 239:487.
4. Saiki, R.K., Bugawan, T.L., Horn, G.T., Mullis, K.B., and Erlich, H.A. (1986) *Nature* 324:163.
5. Gyllensten, U.B., and Erlich, H.A. (1988) *Proc. Natl. Acad. Sci. USA* 85:7652.
6. Scharf, S., Friedmann, A., Steinman, L., Brautbar, C., and Erlich, H.A. (1989) *Proc. Natl. Acad. Sci. USA*, in press.
7. Saiki, R.K., Chang, C.-A., Levenson, C.H., Warren, T.C., Boehm, C.D., Kazazian, Jr., H.H., and Erlich, H.A. (1988) *New Engl. J. Med.* 319:537.
8. Bugawan, T.L., Saiki, R.K., Levenson, C.H., Watson, R.M., and Erlich, H.A. (1988) *Bio/Technology* 6:943.
9. Saiki, R., Walsh, P.S., Levenson, C.H., and Erlich, H.A. (1989) *Proc. Natl. Acad. Sci. USA*, in press.
10. Gyllensten, U., and Erlich, H.A. (1989) *Proc. Natl. Acad. Sci. USA*, in press.
11. Scharf, S.J., Friedmann, A., Brautbar, C., Szafer, F., Steinman, L., Horn, G., Gyllensten, U., and Erlich, H.A. (1988) *Proc. Natl. Acad. Sci. USA* 85:3504.

208 HLA Class II Gene Polymorphism

12. Horn, G.T., Bugawan, T.L., Long, C., and Erlich, H.A. (1988) *Proc. Natl. Acad. Sci. USA* 85:6012.
13. Todd, J.A., Bell, J.I., and McDevitt, H.O. (1987) *Nature* 329:599-604.
14. Sinha, A.A., Brautbar, C., Szafer, F., Friedmann, A., Tzfoni, E., Todd, J.A., Bell, J.I., and McDevitt, H.O. (1988) *Science* 239:1026-1029.
15. Bugawan, T.L., Horn, G.T., Long, C.M., Mickelson, E., Hansen, J.A., Ferrara, G.B., Angelini, G., and Erlich, H.A. (1988b) *J. Immunol.* 141:4024.
16. Bugawan, T.L., Angelini, G., Larrick, J., Auricchio, S., Ferrara, G.B., and Erlich, H.A. (1989) *Nature*, in press.
17. Begovich, A., Bugawan, T.L., Nepom, B.S., Klitz, W., Nepom, G.T., and Erlich, H.A. (1989) Submitted.
18. Erlich, H.A. (1989) *Nature* 337:415.
19. Nomenclature for Factors of the HLA System (1988) *Immunogenetics* 28:391-398

CHAPTER 17

Applications of PCR to the Analysis of Biological Evidence

Cecilia H. von Beroldingen, Edward T. Blake,

Russell Higuchi, George F. Sensabaugh, and Henry Erlich

INTRODUCTION

The ability to detect DNA polymorphisms in biological evidence samples has revolutionized forensic biology. Using restriction fragment length polymorphism (RFLP) analysis, individual-specific DNA "fingerprints" can be obtained, provided that the DNA is relatively undegraded and present in sufficient quantity. At least 50 ng of high molecular weight DNA is required for RFLP analysis using a single locus probe[1] and greater than 100 ng for multilocus probe analysis.[2] However, these amounts of undegraded DNA are frequently not available from forensic evidence in practice. For this reason, some forensic cases involving biological evidence are not amenable to RFLP analysis. Moreover, RFLP consumes a large amount of material and so many samples cannot be re-analyzed.

These limitations may be overcome by employing the polymerase chain reaction (PCR) to amplify specific polymorphic DNA sequences.[3] Use of the thermostable *Taq* DNA polymerase has markedly improved the specificity and yield of the polymerase chain reaction and has made the procedure amenable to automation.[4]   As a result, it is now possible to determine the DNA type of a single hair, a single diploid cell, or even a single sperm.[5,6]   Also, because PCR can generate a large number of copies of a specific DNA sequence, methods used to detect DNA polymorphisms are simpler and less time-consuming than those used in RFLP analysis.

We will describe the use of the polymerase chain reaction in DNA typing of forensic samples and will address the special concerns associated with the handling and analysis of biological evidence.

DEVELOPMENT OF MARKER SYSTEMS FOR PCR

The ultimate goal of forensic DNA analysis is to obtain a positive identification of the donor of a biological evidence sample.   Toward this end, the analysis should employ highly informative genetic markers and be capable of being carried out easily in the typical forensic laboratory.

In order to be of maximum benefit to the forensic scientist, a genetic marker system for forensic PCR analysis should satisfy the following criteria.

1.   The marker should be highly polymorphic and have a high level of genetic heterozygosity.

2.   The target sequence should be easily and specifically amplified.

3.   Methods for detecting allelic variation should be uncomplicated and thoroughly reliable.

4.   Population data on genotype frequencies must be available in order to assign estimates of the marker's power of discrimination and the probability of false inclusion.[20,21]

5.   The marker systems should be inherited independently so that frequencies derived from one marker system can be multiplied with those from others, thereby increasing the power of discrimination. Independent inheritance occurs when the markers are on separate chromosomes or are in linkage equilibrium when present on the same chromosome.

At present, only a few marker systems based on PCR are sufficiently well characterized for use in forensic DNA analysis.   As a result, the power of discrimination of these systems is not as high as those employed in RFLP analysis.   However, there are many polymorphic regions of human DNA that can be exploited by PCR analysis methods so that the level of discrimination possible should soon improve.

DNA polymorphisms can be divided into two categories: sequence polymorphisms, such as occur within the genes of the highly polymorphic HLA complex[7] and the D-loop region of mitochondrial DNA[8]; and length polymorphism, exemplified by the VNTR (variable number of tandem repeat) loci.[9] In both cases PCR may be used to accumulate DNA fragments containing the polymorphic region. Thereafter, different detection strategies are utilized to distinguish allelic variants.[5]

The most straightforward, albeit laborious, method of detecting sequence polymorphism is to determine the DNA sequence of the amplified product using standard sequencing procedures[10-13] (see Chapter 5 of this book). In cases where only a single sequence variant is present, e.g., in mitochondrial DNA, direct sequencing of the PCR product would be unambiguous. In heterozygous individuals containing two allelic variants, directly sequencing the PCR product would yield a mixture of two sequences. Identifying the two different sequences without DNA cloning can be problematic, especially if the two alleles differ in sequence at multiple positions.

Sequence polymorphism can be detected more simply and rapidly using allele-specific oligonucleotide (ASO) probes.[14] Under the appropriate conditions, ASO probes will only hybridize to those sequences to which they are perfectly matched. In order to design ASO probes, the DNA sequence of all the commonly occurring allelic variants must first be determined. Once this has been established, ASO probes can be designed to detect these alleles by differential hybridization. See Chapter 16 (this book) for a complete discussion of ASO detection formats. For our purposes, there are two formats; one in which PCR product is spotted on membranes and challenged with ASO probes,[14,15] and one where the probes themselves are immobilized on a single membrane and used to capture complementary PCR product.[16] In both methods, hybrids can be detected *in situ* by the formation of a colored precipitate.

Length variation among alleles at a given VNTR locus can be readily detected by size fractionation of PCR products on an analytical gel.[17] These products are made using primers that flank the repeat region of the locus. As long as the specificity of the reaction is high and no spurious products are generated, the size of the products should reflect the number of tandem repeats contained within each allele. If the yield and/or specificity of amplification is low, VNTR loci may be detected by blotting and hybridizing with a probe specific for the tandem repeat sequence.[18]

The first and most well-developed system for PCR analysis of forensic samples is the DQα system. The DQα gene is a class II HLA gene which contains a region of allelic variation within the second exon. Using primers that anneal to conserved sequences flanking this region, a 242-bp amplification product is obtained. Thus far, eight alleles have been demonstrated by cloning and sequencing amplification products from various serologically defined cell lines.[11] Oligonucleotide probes have been designed which define the six most common alleles. Four of the probes hybridize to the same sequence variant region of the amplified product and distinguish the A1, A2, A3, and A4 alleles. The A1 allele may be further subtyped as A1.1, A1.2, or A1.3 by using another set of ASO probes which bind to a different sequence variant region upstream from the first.[5] DQα typing of DNA can be simply and rapidly accomplished by dot-blot hybridization in either of the two formats described above. Allele and genotype frequency data have been accumulated for Caucasian, Black, and Asian populations.[19]

The six alleles define twenty-one genotypes with frequencies ranging from less than 0.005 to 0.15. The discriminating power (DP)[20] of the DQα typing system, i.e., the probability of distinguishing between two individuals chosen at random from these populations together, is 0.93. This compares favorably with the discriminating power of traditional genetic marker systems such as ABO blood groups (DP=0.60) and the isoenzyme PGM (DP=0.76).[21] Although far from approaching individual identification, the DQα system has proved useful in excluding or including suspects in criminal cases where conventional genetic typing has failed or where insufficient DNA was available for RFLP analysis. DQα typing has been used in forensic casework for more than two years (see example below).

Another typing system utilizing polymorphisms in the HLA-D region is based upon sequence variation at the DPß locus. So far, 21 allelic variants have been identified within the highly polymorphic second exon of the DPß gene based on PCR-derived sequence information.[22,23] The variant sequences are largely localized to six regions and can be distinguished by a panel of 14 sequence-specific oligonucleotide (SSO) probes. Typing of amplified DPß sequences is carried out in a dot-blot format, just as with DQα, but the interpretation of the probe hybridization pattern is more complex. This is because few of the probes are specific for a particular allele; rather, different combinations of probes bind to different alleles. Nevertheless, DPß can be used as a highly informative marker system for individual identification. Furthermore, preliminary results indicate that DPß alleles are in linkage equilibrium with DQα alleles,[24] thereby allowing the power of discrimination to increase multiplicatively when these systems are used in combination.

Another potential marker system is the D-loop region of human mitochondrial DNA. Sequence analysis of this region indicates a high degree of variability.[7] Sequence variation may be detected either by direct sequencing or by dot-blot hybridization with a panel of SSO probes.

An advantage of mitochondrial DNA is the 10-10,000 fold higher copy number (depending on the tissue) of mitochondrial sequences compared to single copy genomic sequences. When dealing with limited amounts of DNA, the likelihood of amplifying a mitochondrial sequence is correspondingly higher.

Recently, Jeffreys et al.[18] have reported the amplification of VNTR alleles up to about 10 kb in length. At least six different VNTR loci may be coamplified from nanogram amounts of human DNA and are simultaneously detected by Southern blot hybridization of the products with a battery of locus-specific probes. The resulting DNA fingerprint has a degree of individual specificity approaching 10^{-5}. Multilocus amplification was also obtained from single human cells, although artifactual "alleles" were sometimes generated.

Several factors affect the efficiency and accuracy of VNTR amplifications.[18] There is an inverse relationship between the size of the VNTR allele and the efficiency of amplification: the shorter the allele, the more efficient the amplification. Since the length of VNTR alleles at a particular locus can be quite disparate, the longest alleles tend not to be detected. In ref. 18, increasing extension time increased the product yield for longer alleles. However, there still remains a significant molar imbalance between short and long allele products. Increasing cycle number can exacerbate this imbalance and can also generate spurious amplification products such as single-stranded products, partially extended products, or bogus alleles arising from out-of-register annealing of VNTR templates. This latter phenomenon becomes evident when there is a high concentration of PCR product in the reaction.

Theoretically, the most faithful and efficient amplification of VNTR alleles should be obtained at loci where the alleles are not widely disparate in length and where the amplified products are relatively short (<1 kb). In this case, all the alleles would be amplified with comparable efficiencies. For certain VNTR loci, it has proven possible to visualize allelic variants on an ethidium-stained agarose gel.[17] For other loci, Southern blot hybridization may still be necessary to detect VNTR alleles. The use of nonradioactive probes, however, would simplify such an analysis, making it adaptable to the typical forensic laboratory (Scharf and Erlich, unpublished).

Finally, there are a number of dimorphic loci with allele frequencies approaching 50% that are amenable to PCR analysis in a dot-blot format, such as the low density lipoprotein receptor (LDLr) gene.[25] Simultaneous amplification and typing of many of these loci, each on different chromosomes, can be a system with good discrimination power (Saiki and Erlich, pers. comm.). Gibbs *et al.* (see Chapter 15, this book) have amplified up to nine different loci at once. A Y chromosome-specific marker can also be amplified by PCR.[26] The male-specific product can be directly visualized on an ethidium bromide-stained gel or be analyzed in a dot-blot format.

In summary, a number of highly informative PCR-based typing systems are under development which should make PCR analysis more useful in individual identification. Nevertheless, given that many evidence samples are not suitable for RFLP analysis, we feel that the existing PCR systems will already be of much use. Furthermore, the detection methods used in PCR analysis are rapid, uncomplicated, and do not require the use of radioisotopes. Typing methods involving hybridization with sequence specific oligonucleotide probes also confer the advantage that alleles are defined qualitatively in a plus or minus manner. Therefore, the assembly and use of population data bases is simplified.

THE POWER OF PCR AND SPECIAL CONCERNS IN THE ANALYSIS OF BIOLOGICAL EVIDENCE

All forms of biological evidence typically collected during the examination of a crime scene or an individual are amenable to analysis by PCR. Thus it has been demonstrated that DNA typing results can be obtained from semen stains, blood stains, single hairs, and buccal epithelial cells.[18,27] In theory, even minute amounts of degraded DNA are typeable, as long as there is a single intact DNA strand spanning the target sequence being amplified.

An example of the power of PCR is illustrated by the ability to type DQα DNA amplified from formaldehyde-preserved autopsy samples in which the DNA was profoundly degraded.[28] In this circumstance, the average DNA size was only about 100 bp, too small to allow efficient amplification of the 242-bp DQα target sequence. Instead, primer pairs were chosen that amplified 166-bp and 82-bp polymorphic sites at this locus. These fragments were then typed by ASO probe/dot-blot hybridization. PCR has also been applied to the analysis of degraded DNA found in a 7000-year old human brain.[29] Sequence analysis was performed on amplified mitochondrial DNA. Again, the DNA was so degraded that only regions less than 200 bp in length could be successfully amplified.

In addition to degradation, DNA in evidence samples may undergo modifications either spontaneously or as a result of exposure to UV irradiation or chemical agents. These lesions could affect the ability of the DNA to be amplified efficiently. It is not yet known for certain whether *Taq* polymerase is prone to misincorporation errors when it encounters damaged templates or whether incorporation is simply blocked. Evidence to date indicates that DNA damage decreases amplification product yield but does not lead to artifactual typing results, suggesting that undamaged DNA serves as a preferential template for amplification. Thus, for example, UV irradiation of template DNA decreases the level of amplification but the product has the same DNA type as undamaged DNA.[30] In addition, amplification of ancient DNAs known to contain a variety of lesions generates unambiguous sequence data.[31]

Another factor influencing the ability of evidence DNA to be typed is the presence of inhibitory substances that copurify with DNA. Blood stains in particular are sometimes refractory to amplification. Such inhibitory activities have also been encountered in amplifying ancient DNAs.[29] These inhibitory activities may be mitigated in some cases by adding additional *Taq* polymerase or by diluting the DNA sample before amplification.

The remarkable sensitivity of PCR is the procedure's main advantage over RFLP analysis, but it is also the source of a potential concern, that of contamination. Foreign DNAs may derive from several different sources and may become associated with the sample before the evidence is collected, during evidence handling, or during sample analysis. By exercising reasonable precautions and being attuned to signs of possible contamination, however, problems may be anticipated and avoided.[32]

Although bacterial and fungal DNAs may contribute to the DNA isolated from an evidence sample, these DNAs would not amplify using primers designed to amplify human sequences. A more pressing concern is the presence of heterologous DNA of human origin. In some cases, the presence of a DNA mixture is unavoidable. For example, many semen samples collected in sexual assault cases contain DNAs from two sources:

epithelial cells from the victim and sperm from the rapist. These DNAs may be separated by a differential cell lysis procedure that takes advantage of the insensitivity of sperm to lysis in the absence of a reducing agent.[33] Depending upon the circumstances of the crime, other DNA mixtures are possible, including any combination of blood, saliva, semen, hair, or other tissues. Some of these mixtures are not separable by differential lysis. However, prior to DNA analysis, careful evidence examination should reveal many of these situations and allow a more informed interpretation of the typing data.

The most likely avenue for the chance introduction of a foreign DNA is during sample preparation. The foreign DNA may arise from cross-contamination between samples, or from contamination of a sample with PCR product from some previous reaction. By undertaking certain common sense precautions, the occurrence of adventitious contamination can be minimized and eliminated.

Isolation of DNA from evidence samples, the setting up of the amplification reactions, and the analysis of the PCR product should be performed in physically separate laboratory areas. In preparing DNA for subsequent amplification, it is imperative that samples likely to contain significant amounts of DNA, such as from whole blood reference samples or large semen stains, be treated separately from samples that contain minute amounts of DNA, such as single hairs. Amplification reactions should be set up in another area or within a laminar flow hood to prevent possible airborne carryover of PCR product.

The potential problem of PCR product carry-over contamination is one that is endemic to any research, clinical, or forensic laboratory using PCR. Consider the fact that an amplification reaction will frequently generate microgram amounts of a specific product. One-tenth μl of such a reaction introduced inadvertently by a contaminated pipettor barrel could introduce as many as 10^9 templates into a new reaction. As a consequence, amplification of the contaminant could make typing results ambiguous or even mask the type of the sample DNA.

Sterile plasticware, sterile reagent solutions, and disposable gloves should be used throughout the preparation and analysis of evidence DNA. Separate sets of pipetting devices for each work area are advisable, some limited to reagent use only, others (preferably of the positive displacement type with disposable tip and plunger) for pipetting sample DNAs, and still another set for analysis of PCR products.

Fortunately, the presence of a contaminant is often readily apparent. In a dot-blot hybridization assay, more than two alleles may be detected. In a sequence analysis, contaminants would give rise to ambiguities in the sequence data. For a VNTR amplification, more than the expected number of bands would be visible on the analytical gel or Southern blot. More problematical is discounting the possibility of a chance contaminant being preferentially amplified and masking the evidence type or giving a result when there was actually no amplifiable DNA present in the sample. Appropriate controls should be used that would indicate these situations.

Negative controls, including extraction reagent blanks and unstained substrate controls (when possible), should be run for each batch of DNA isolations. Likewise, each set of amplification reactions should include negative reagent controls. The absence of a PCR product in these control reactions attests to the validity of the typing results derived from the evidence samples.

It is desirable to have some form of redundant sample analysis. Because of the sensitivity of PCR, one has the luxury of being able to make more than a single DNA preparation from the same evidence without exhausting the entire sample. Furthermore, only a portion of the isolated DNA is used for each amplification. Thus, if the duplicate samples fail to yield identical results, one has the option of reamplifying from the same DNA samples or going back to the original piece of evidence to carry out a new DNA isolation. Duplicate analyses yielding identical DNA types also serve to independently verify results, refuting the objection that a sporadic contaminant may have been amplified.

CASEWORK EXAMPLE

The following example illustrates the application of PCR to the analysis of evidence from a sexual assault case in which the victim had had sexual intercourse with her husband nine hours before the alleged assault. Evidence was submitted to determine whether the Caucasian suspect could be eliminated as the sperm donor. Vaginal epithelial cell DNA and sperm DNA were isolated from two partial vaginal swabs and from cell debris derived from vaginal swabs by a differential lysis procedure. In the first step, epithelial cells were preferentially lysed in a digestion buffer containing detergent and protease.[33] The sperm heads, which are resistant to digestion in the absence of a reducing agent, are collected by centrifugation, repeatedly washed, and resuspended in digestion buffer to which dithiothreitol (DTT) has been added. The effectiveness of the differential lysis procedure is demonstrated in Figure 1, which shows the cellular material collected from one of the vaginal swabs before and after the first digestion step. Before digestion, many epithelial cells as well as sperm are visible whereas predominantly sperm heads are present in the pellet recovered after digestion.

DNAs from epithelial cell fractions, sperm fractions, and whole blood samples from the victim, victim's husband, and suspect were amplified and the DQα types determined by hybridization with ASO probes in a dot-blot format. Figure 2 shows the DQα type of the victim to be 3,4, that of the husband 1,2, and that of the suspect 1,1. DNAs from homozygous typing cell lines are included as typing standards. The sperm DNA recovered from both partial vaginal swabs and from the cell debris sample typed as 1,1, indicating a clean separation of sperm DNA from vaginal epithelial cell DNA. The epithelial cell fractions gave a mixture of alleles 1, 3, and 4, consistent with the presence of some sperm DNA in the epithelial cell

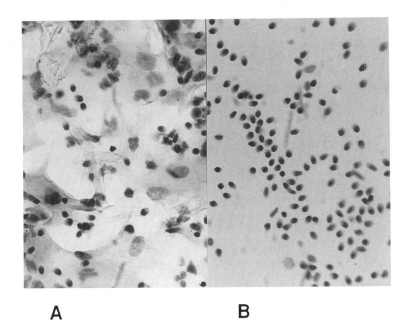

A   B

Figure 1. Differential cell lysis of a semen sample. **A.** A small aliquot of cellular material collected from vaginal swab #3 was stained with nuclear fast red and picroindigo carmine. Both intact spermatozoa and epithelial cells are observed. **B.** After digestion in the absence of DTT, the resistant cellular material was pelleted and an aliquot removed and stained as above. Almost entirely sperm heads are observed.

fraction. (This is a common occurrence when there are a large number of sperm in a sample.) Subtyping of DNAs containing the DQA1 alleles (Figure 3) revealed the DQα type of the sperm donor to be either 1.2,1.3, or 1.3,1.3. The ASO subtyping probes used in this format do not distinguish between these two genotypes. The suspect's DNA gave the same result. Figure 4 shows the DQα typing results obtained using the immobilized probe strip format on DNA from the victim, husband, and suspect, and from the three sperm fractions. An additional probe on the strip allows one to distinguish between a 1.2,1.3 type and a 1.3,1.3 type. The victim's DNA types as 3,4 and that of the husband as 1.2,2. The suspect's DNA types as 1.2,1.3, the same type as the sperm donor. Trace amounts of allele 2 were observed in amplifications from the sperm fractions. The presence of this allele is consistent with the fact that the

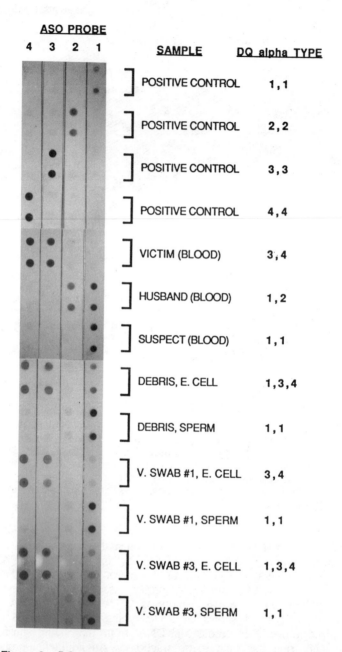

Figure 2. DQα typing of evidence samples. DNAs from the samples described in the text and from homozygous typing cell lines were amplified and the PCR product spotted in replicate onto nylon membranes. The membranes were hybridized separately with one of four nonradioactive probes that recognize alleles A1, A2, A3, and A4. Hybridization was visualized by the formation of a colored precipitate.

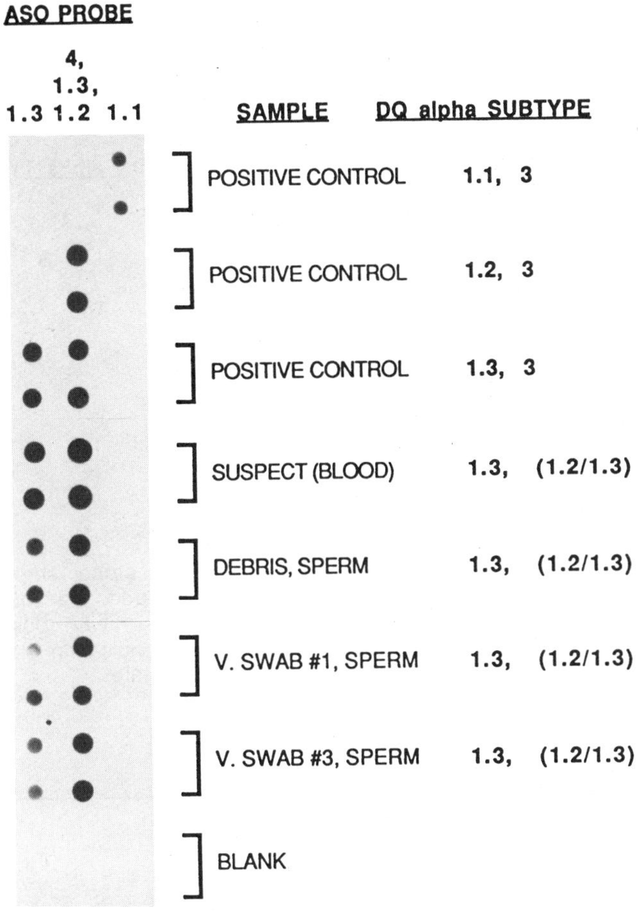

ASO PROBE

```
4,
1.3,
1.3 1.2 1.1        SAMPLE    DQ alpha SUBTYPE
```

SAMPLE	DQ alpha SUBTYPE
POSITIVE CONTROL	1.1, 3
POSITIVE CONTROL	1.2, 3
POSITIVE CONTROL	1.3, 3
SUSPECT (BLOOD)	1.3, (1.2/1.3)
DEBRIS, SPERM	1.3, (1.2/1.3)
V. SWAB #1, SPERM	1.3, (1.2/1.3)
V. SWAB #3, SPERM	1.3, (1.2/1.3)
BLANK	

Figure 3. DQα subtyping of evidence samples. DNAs amplified from the suspect's blood, the three sperm fractions, and typing controls were immobilized in replicate on a nylon membrane and challenged with radioactive ASO probes that distinguish the A1 subtypes. Hybrids were detected by autoradiography.

victim had intercourse with her husband earlier on the day of the incident. The reagent blanks in both the typing assays gave negative results (Figures 3 and 4). DQα type 1.2,1.3 occurs in 2.6% of the Caucasian population. Thus, the suspect and 2.6% of the male Caucasian population cannot be eliminated as the sperm donor in this case.

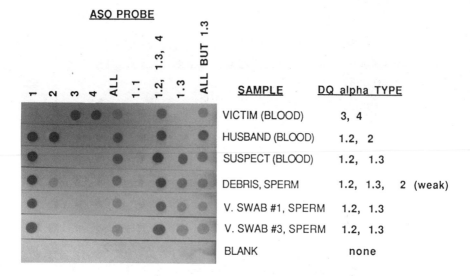

Figure 4.   DQα typing using immobilized probe strips. DNAs amplified from reference samples and sperm fractions were hybridized to DQα ASO probes that had been immobilized on a nylon membrane strip. Hybridization was visualized by the formation of a colored precipitate.

PROTOCOLS

DNA Extraction Procedures for Biological Evidence

I. Stock Solutions

- 1 M Tris.HCl, pH 8.0
- 0.5 M EDTA, pH 8.0
- 5 M NaCl
- 10% SDS (sodium dodecyl sulfate)
- 1 M DTT (dithiothreitol) in 10 mM sodium acetate, pH 5.2 stored frozen
- Proteinase K, 10 mg/ml in H_2O, aliquoted and stored frozen
- 25:24:1 Phenol/Chloroform/Isoamyl alcohol [Note: phenol must be buffered by washing several times with 50 mM Tris.HCl, pH 8.0 and stored under TE buffer (10 mM Tris.HCl, 1 mM EDTA, pH 7.5)]

- Digestion Buffer: 10 mM Tris.HCl, pH 8.0, 10 mM EDTA, 50 mM NaCl, 2% SDS
- Water-saturated n-butanol

II. Digestion Procedures

A. Blood Digestion

- Suspend buffy coat, 0.1 ml whole blood, or blood stain in 0.5 ml digestion buffer containing 15 μl proteinase K solution.   Incubate at least 1 h at 56°C.   Remove blood stain substrate.
- Add 0.5 ml phenol/chloroform/isoamyl alcohol.   Vortex.   Spin in microcentrifuge 2 min.
- Transfer top (aqueous) phase to new tube.   Repeat extraction until interface is clear.
- Remove residual phenol/chloroform by extracting with an equal volume of n-butanol.
- Remove lower layer for Centricon 30 concentration.

B. Sperm Digestion in the Presence of Contaminating Cells

- Suspend sample in 1 ml sterile deionized water.   Incubate at RT for 30 min to elute cellular material.   Vortex gently.   Remove substrate.   Spin 1 min in microcentrifuge, discard supernatant.
- Remove a small aliquot of the pellet for microscopic examination.
- Suspend pellet in digestion buffer and treat with proteinase K as described above.   Digest for no more than 2 hr.   Spin in microcentrifuge 1 min to pellet sperm heads.   Remove supernatant for epithelial cell DNA analysis (or discard).
- Wash sperm pellet with 0.5 ml digestion buffer.   Spin 1 min in microcentrifuge, discard supernatant.   Repeat.   (Wash steps are critical, particularly when the ratio of sperm to epithelial cells is small.)
- Wash once with water.   Remove small aliquot of pellet for microscopic examination to verify digestion of epithelial cells and recovery of sperm.
- Suspend pellet in 0.5 ml digestion buffer containing 20 μl DTT and 15 μl proteinase K.   Incubate for at least 1 hr at 56°C.
- Extract DNA as above.

C. Hair Digestion

- For found hairs, wash once with sterile deionized water.   For mounted hairs, freeze slide and remove cover slip with scalpel.   Wash away mounting medium with xylene.   Wash hair with ethanol, then with sterile deionized water.
- Incubate in 0.5 ml digestion buffer containing 20 μl DTT and 15 μl proteinase K for several hours at 56°C.   Add additional aliquots of DTT and proteinase K if necessary to achieve complete digestion.

- Spin in microcentrifuge briefly to remove pigment.
- Extract DNA as above.

III. Centricon 100 Concentration

DNA from evidence samples is routinely purified and concentrated using a Centricon 100 ultrafiltration device (Amicon) to maximize DNA recovery.
- Add 1.5 ml TE buffer to the Centricon column.   Add butanol-extracted DNA (~0.5 ml).

- Centrifuge at 20°C in a fixed angle rotor for 20 min at 5000 rpm.

- Add 2 ml TE buffer to concentrated DNA and centrifuge.   Repeat.

- Collect the concentrated DNA in the tube provided by brief centrifugation. The volume is ~25-40 µl.

A 5-10 µl aliquot of each sample is electrophoresed on a 1% agarose gel to estimate DNA quantity and quality.   Up to half the remaining sample is used for amplification.

ACKNOWLEDGMENTS

We gratefully acknowledge the helpful comments of Linton von Beroldingon and Sean Walsh.

REFERENCES

1. Wong, Z., Wilson, V., Patel, I., Povey, S., and Jeffreys, A.J. (1987) *Ann. Hum. Genet.* 51:269-288.
2. Jeffreys, A.J., Wilson, V., and Thein, S.L. (1985) *Nature* 316:76-79.
3. Mullis, K.B., and Faloona, F. (1987) *Meth. Enzymol.* 155:335-350.
4. Saiki, R.K., Gelfand, D.H., Stoffel, S., Scharf, S.J., Higuchi, R., Horn, G.T., Mullis, K.B., and Erlich, H.A. (1988) *Science* 239:487-494.
5. Higuchi, R., von Beroldingen, C.H., Sensabaugh, G.F., and Erlich, H.A. (1988) *Nature* 332:543-546.
6. Li, H., Gyllensten, U.B., Cui, X., Saiki, R.K., Erlich, H.A., and Arnheim, N. (1988) *Nature* 355:414-417.
7. Trowsdale, J., Young, J.A.T., Kelly, A.P., Austin, P.J., Carson, S., Meunier, H., So, A., Erlich, H.A., Speilman, R.S., Bodmer, J., and Bodmer, W.F. (1985) *Immunol. Reviews* 85:5-43.
8. Aquadro, C.F., and Greenberg, B.D. (1983) *Genetics* 103:287-312.
9. Nakamura, Y., Leppert, M., O'Connell, P., Wolff, R., Holm, T., Culver, M., Martin, C., Fujimoto, E., Hoff, M., Kumlin, E., and White, R. (1987) *Science* 235:1616-1622.
10. Scharf, S.J., Horn, G.T., and Erlich, H.A. (1986) *Science* 233:1076-1078.
11. Horn, G.T., Bugawan, T.L., Long, C.M., and Erlich, H.A. (1988) *Proc. Natl. Acad. Sci. USA* 85:6012-6016.
12. Wrischnik, L.A., Higuchi, R.H., Stoneking, M., Erhlich, H.A., Arnheim, N., and Wilson, A.C. (1987) *Nucl. Acids Res.* 15:529-535.
13. Gyllensten, U.B., and Erlich, H.A. (1988) *Proc. Natl. Acad. Sci. USA* 85:7652-7656.
14. Saiki, R.K., Bugawan, T.L., Horn, G.T., Mullis, K.B., and Erlich, H.A. (1986) *Nature* 324:1350-1354.

15. Saiki, R.K., Chang, C.-A., Levenson, C.H., Warren, T.C., Boehm, C.D., Kazazian, Jr., H.H., and Erlich, H.A. (1988) *N. Engl. J. Med.* 319:537.
16. Saiki, R.K., Walsh, P.S., Levenson, C.H., and Erlich, H.A. (1989) *Proc. Natl. Acad. Sci. USA*, in press.
17. Boerwinkle, E., Xiong, W., Fourest, E., and Chan, L. (1989) *Proc. Natl. Acad. Sci. USA* 86:212-216.
18. Jeffreys, A.J., Wilson, V., Newman, R., and Keyle, J. (1988) *Nucl. Acids Res.* 16:10953-10971.
19. Madej, R., Helmuth, R., Louie, P., Hom, G.T., and Blake, E.T. Unpublished data.
20. Fisher, R.A. (1951) *Heredity* 5:95-102.
21. Sensabaugh, G.F. (1982) in *Handbook of Forensic Sciences*, Saferstein, R., ed. Prentice Hall, pp. 338-415.
22. Bugawan, T.L., Hom, G.T., Long, C.M., Mickelson, E., Hansen, J.A., Ferrara, G.B., Angelini, G., and Erlich, H.A. (1988) *J. Immunol.* 141:4024-4030.
23. Begovich, A.B., Bugawan, T.L., Nepom, B., Klitz, W., Nepom, G.T., and Erlich, H.A. (1989) Submitted.
24. Bugawan, T.L., and Erlich, H.A. (1988) Unpublished data.
25. Hobbs, H.H., Esser, V., and Russel, D.W. (1987) *Nucl. Acids Res.* 15:379.
26. Kogan, S.C., Doherty, M., and Gitschier, J. (1987) *N. Engl. J. Med.* 317:985-990.
27. Higuchi, R., and Blake, E.T. (1989) in *DNA Technology and Forensic Science*. Banbury Report #32. J. Ballantyne, G. Sensabaugh, J. Witkowski, eds. Cold Spring Harbor Press, in press.
28. Bugawan, T.L., Saiki, R.K., Levenson, C.H., Watson, R.M., and Erlich, H.A. (1988) *Biotechnology* 6:943-947.
29. Pääbo, S., Gifford, J.A., and Wilson, A.C. (1988) *Nucl. Acids Res.* 16:9775-9786.
30. Buoncristiani, M., von Beroldingen, C.H., and Sensabaugh, G.F. (1989) Unpublished results.
31. Pääbo, S. (1989) *Proc. Natl. Acad. Sci. USA* 86:1939-1943.
32. Kwok, S., and Higuchi, R. (1989) *Nature*, in press.
33. Guisti, A., Baird, M., Pasquale, S., Balazs, I., and Glassberg, J. (1986) *J. For. Sci.* 31:409-417.

CHAPTER 18

Detection of *ras* Oncogenes Using PCR

Johannes L. Bos

INTRODUCTION

Cancer is thought to be mainly due to alterations in the cellular genome that affect the expression or function of genes controlling cell growth and differentiation. Present-day molecular cancer research aims at identifying the alterations responsible for the development of tumors, at characterizing the genes involved and at determining the consequences of the gene alterations for the control of cell growth and differentiation and for the process of carcinogenesis. Five major approaches have been used to identify genes that may place a role in carcinogenesis: 1) characterization of cellular sequences transduced to the genome of acutely transforming retroviruses, 2) transfection of tumor DNA into mouse NIH/3T3 cells and selection for transforming genes, 3) isolation of genes altered by insertion or viral sequences, 4) direct isolation of altered genes from tumor cells with a visible alteration, such as a translocation in their chromosomes, and 5) isolation of genes with sequence similarities to genes discovered by the other approaches. By these procedures, a large number of genes have been identified, but to date only a fraction of these have been implicated in the development of human tumors. Of these, the *ras* family has been the most thoroughly investigated.

The *ras* family consists of three closely related genes, H*ras*, K*ras*, and N*ras*, which have been characterized as potential transforming genes by their presence in certain acutely transforming retroviruses and in transformed NIH-3T3 cells transfected with DNA isolated from a variety of tumors.[1] These three genes acquire transforming potential when, due to a point mutation, a single amino acid of the protein product at one of the critical positions 12, 13, or 61 is altered. In the years following the initial discovery, many reports appeared identifying transforming *ras* genes in a variety of tumor cell lines and occasionally also in human tumor tissues. These studies revealed that the occurrence of transforming *ras* in human tumors is widespread and disclosed the positions at which the mutations may occur, but for several reasons they contributed little to our understanding of the role of transforming *ras* genes in the development of specific tumor types. The numbers of tumors analyzed were too small, since the NIH/3T3 focus formation assay is too laborious and requires too much high-quality DNA for large-scale analysis. Furthermore, the results obtained with cell lines must be interpreted with caution, since the gene may have incurred the mutations during the establishment of the cell lines, or, alternatively, the cell line may have been derived from a rare aberrant cell in the tumor. The methodological problems have been solved by the introduction of biochemical assay systems. These procedures are based on the direct identification of the point mutations responsible for the transforming properties of the *ras* genes. In one of the procedures, the transforming mutation is identified by specific hybridization to oligodeoxynucleotide (ODN) probes.[2,3] A mutation-specific probe has been synthesized for each possible single-base, non-silent mutation in codons 12, 13 and 61, since it is in these codons so that most, if not all, transforming mutations occur. This implies that 63 different probes are needed for this analysis. The probes are radiolabeled, synthetic icosadeoxynucleotides. The DNA from a tumor is isolated, either from a well characterized part or, preferably, from frozen[4] or paraffin-embedded[5] tissue sections. The percentage of tumor cells in the sample should be at least 25 percent. Segments of the three *ras* genes are selectively amplified approximately one-million-fold by the polymerase chain reaction.[6] Subsequently, the DNA which now consists mainly of *ras*-specific sequences is spotted onto a solid support and hybridized to the mutation-specific probes. Only the probe that matches with the *ras* point mutation will hybridize and provide an autoradiographic signal.

An alternative procedure is the RNAase A mismatch cleavage method.[7,8] DNA, selectively amplified *in vitro* for *ras*-specific sequences, is hybridized to radiolabeled RNA complementary to the unmutated sequence. At positions of mismatch, RNAase A will cleave the RNA resulting in two fragments that can be identified after gel electrophoresis. Recently, *ras* gene mutations were identified by direct sequencing of the *in vitro* amplified material.[9,10]

In the last few years, these biochemical procedures, as well as the NIH/3T3 transfection assay, have been used to detect and identify *ras* mutations in several tumor types. The results of these studies, concerning the incidence of mutated *ras* genes in the various tumor types, is presented in Table 1. For information about earlier studies, I refer to a recent review.[11]

The main conclusions to be drawn from inspecting Table 1 is that some tumor types never, or only very infrequently, harbor a mutated *ras* gene, whereas others have mutated *ras* genes with varying incidences. The highest incidences are found in adenocarcinomas of the pancreas (90%), the colon (50%) and the lung (30%), in thyroid tumors (50%) and in myeloid leukemia (30%).

From several tumors, additional information is available about the presence of the mutated *ras* gene, the stage of disease at which the mutational event occurs, and a possible relation between clinical and histopathological features.

PANCREATIC CARCINOMAS

The highest incidence of *ras* gene mutations has been found in adenocarcinoma of the exocrine pancreas: 21 of the 22 carcinomas were found to have incurred mutations at or around codon 12 of the K*ras* gene using the RNAase cleavage mismatch assay.[8] These mutations were not present in the surrounding normal tissue. With the ODN hybridization assay, 28 of the 20 samples were found to contain a mutated *ras* gene, all in codon 12 of the K*ras* gene.[12] These data imply that mutation of the K*ras* gene is a critical event in the development of pancreatic tumors. The relevance of mutated *ras* genes in the development of pancreatic tumors is further illustrated by transgenic mice harboring a mutated H*ras* transgene under the control of the pancreas-specific elastase promoter.[13] These mice developed carcinomas of the fetal pancreas directly after the onset of elastase gene expression, indicating that the expression of the mutant *ras* transgene is responsible for the development of the tumor.

COLON CARCINOMAS

Mutations in *ras* genes were found in almost 50 percent of a large number of the colon tumors. In two series, 117 samples were analyzed by the ODN hybridization assay, 54 of which were found to have a *ras* mutation.[4,14] In most cases, K*ras* was found to be mutated, in particular in codons 12 and 13, but, infrequently, mutations in N*ras* have been found, as well. With the RNAase mismatch-cleavage method, 66 samples were analyzed for the presence of mutations in K*ras* and 26 samples were found to contain such a mutation.[15] In the larger benign colon, adenomas- or polyps-mutated *ras* genes occur with a similar frequency.[14] These adenomas are considered to be the precursor lesion of carcinomas and thus, it seems

Table 1. Incidence of *ras* Gene Mutations in Human Malignancies

Tumor Type	Incidence[t]	*ras*[‡]	References
breast cancer	>5		38
			39
			40
ovary	<5		41
cervical ca	<5		11
esophageal ca	<5		42
glioblastoma	<5		11
neuroblastoma	<5		43
stomach ca	5		44
			45
kidney ca	10		46
skin keratoacanthoma	10		47
lung			
epidermoid ca	<5		16
large cell ca	<5		16
adeno ca	30	K-*ras*	16,17
colon			
adeno ca	50	K-*ras*	4,15
adenoma	50	K-*ras*	14
pancreas ca	90	K-*ras*	8,12
seminoma	40	K-*ras*, n-*ras*	11
melanoma	20	N-*ras*	21
			22
bladder ca	6		48
			49
			50
thyroid follicular	50	N-, K-, H-*ras*	18
			19
thyroid papillary	<5		20
myeloid			
MDS	30	N-*ras*	28
			35
			36
AML	30	N-*ras*	25,26
			32
CML	10	N-*ras*	29
			11
			51
lymphoid			
ALL	10	N-*ras*	52
			53
			34
CLL	<5		53
NHL	<5		53
			Yunis and Bos, unpublished

[t] percentage of tumors harboring a mutant *ras* gene.
[‡] *ras* gene preferentially activated.

that in most cases the *ras* gene mutation occurs prior to the conversion from adenoma to carcinoma. The timing of the mutational event is not invariant, however. Several heterogeneous tumors, which upon histological examination clearly contained patches of carcinoma tissue surrounded by adenoma tissue, were screened for the presence of *ras* mutations. Although in five out of six cases analyzed, the mutation found was present in both tissues types, in one case the mutation was only present in the carcinoma tissue.[4,14] In smaller adenomas, the incidence of mutated *ras* genes appears to be much lower than in the larger ones (two out of thirteen).[14] These smaller adenomas regress frequently and progress into a carcinoma in only 10% of the cases. This may imply that the presence of a *ras* mutation provides the adenoma with a growth advantage and, consequently, with a greater probability to progress into a malignant carcinoma. A low incidence of *ras* mutations was also observed in small adenomas from patients with familial adenomatous polyposis, a hereditary predisposition to colon cancer (12%).[14]

NON-SMALL CELL LUNG CARCINOMAS

In two studies, a total of 77 NSCLC have been examined using the oligonucleotide hybridization procedure.[16,17] Fourteen K*ras* and one H*ras* mutations, all in codon 12, were detected. These mutations, however, were exclusively in the 45 adenocarcinomas analyzed, showing the high degree of specificity for this type of lung tumors. Histologically and clinically, adenocarcinomas with a *ras* mutation are largely similar to adenocarcinomas without one. The carcinomas with a *ras* mutation seem slightly smaller and metastasize at a slightly later stage, but the number of tumors analyzed was too small to allow a firm conclusion on that point. There is, however, a positive relation between the smoking history of the patients and the presence of a mutant *ras* gene, suggesting that a mutagenic component in tobacco smoke may cause the *ras* mutation.[16,17]

THYROID CARCINOMAS

The tumors of the follicular epithelium of the thyroid gland represent various stages of tumor progression. The benign (micro-) follicular adenomas are considered to progress at low frequency to follicular carcinomas and, subsequently, to undifferentiated carcinomas. Papillary carcinomas, which are clearly distinguishable from follicular carcinomas, may also progress to undifferentiation carcinomas. Mutations in all three *ras* genes were found both in the benign follicular adenomas, as well as in the follicular and undifferentiated carcinomas, in half of the cases.[18,19] Apparently, the *ras* mutation can occur early in the development of this tumor type. In papillary carcinomas, a low incidence of mutations is observed[20] and no *ras* mutations were detected in macrofollicular hyperplasias.   However, these hyperplasias progress hardly ever, if at all, into malignant tumors.[18]

MELANOMAS

Melanomas are highly malignant tumors which metastasize very rapidly. The primary tumors are usually small and in some cases even occult. In a study using the NIH/3T3 focus formation assay comprising thirteen melanomas, both primary tumors and metastases, only one activated N*ras* gene was found in a metastasis.[21] By ODN hybridization, seven of thirty-five patients with primary tumors or metastases were found to harbor a mutated N*ras* gene.[22] In all four cases where it could be analyzed, the *ras* mutation was present both in the primary tumor and the metastasis.   In two patients, the primary tumor contained two mutated *ras* alleles, whereas the corresponding metastases harbor only one of the two mutated alleles.   In one case, different metastases contained either mutated allele, in the other case, only one mutated allele was found in the metastases.   This suggests that in these two cases, the primary tumor consists of two cell populations, each with a different *ras* mutation.

MYELOID LEUKEMIA

The most interesting type of cancer with respect to *ras* gene mutations, is myeloid leukemia. Myeloid leukemia comprises a large number of diseases, which can be divided into two main groups.   First, myelodysplastic syndrome (MDS) or preleukemic, and secondly, acute myeloid leukemia (AML).   MDS is a heterogeneous group of disorders characterized by abnormally low counts of one or more of the blood lineages, combined with bone marrow abnormalities that may evolve into AML in one-third of the cases.   By cytogenetic analysis and X-chromosome inactivation studies, it has been shown that the initiating event in this disease most likely affects an early stem cell.[23]   The subsequent progression of MDS to AML is associated with a gradual increase in blast cells, which reflects the outgrowth of a new leukemic cell clone.[24]   Arbitrarily, patients with more than 30% immature blast cells in the bone marrow are diagnosed as AML. For most patients with AML, it is unknown whether they have had prior MDS.   In about one-third of the MDS and AML patients, a mutated *ras* gene is detected, mostly in the N*ras* gene, but also in the K*ras* and, infrequently, in the H*ras* gene.[25-36]   In some MDS patients, the mutated *ras* gene is readily detectable in the bone marrow as well as in the peripheral blood cells, even in the mature T-lymphocytes,[29,36] whereas in other patients, the mutation is present in a subpopulation of the bone marrow cells, most likely the leukemic blast cells.[28,36]   Apparently, with respect to the stage at which *ras* mutations occur, patients can be divided into at least two groups.[36]   The first group has incurred a mutated *ras* gene in a multipotent stem cell and, therefore, the mutation is present in all bone marrow cells as well as in the mature peripheral blood cells. In these patients, cells with a *ras* mutation may persist after complete clinical remission.[36] The second group of patients has obtained a *ras* mutation later in the course of the disease and the cells harboring a *ras* mutation represent a new evolving leukemic cell clone. In these patients, the mutant-*ras*-containing cells may be eradicated by treatment and may not reappear after clinical relapse.[32,36,37]

In MDS and AML, the presence of *ras* mutations might have some direct clinical relevance. Firstly, as pointed out by several investigators, MDS patients with a *ras* gene mutation have a higher chance to progress into AML and have poorer prognosis.[28,30,35,36] However, some MDS patients with a *ras* mutation have a stable disease,[33,35] indicating that a *ras* mutation is not a single prognostic factor. Secondly, the *ras* mutation might be a suitable marker for monitoring the effect of chemotherapy and detecting minimal residual disease.[32,34,36]

CONCLUDING REMARKS

From the above summary, it is clear that the PCR procedure has greatly simplified the analysis of *ras* mutations in human tumors, so that it has been possible to survey a large number of different tumor types at a sensitivity that could not be achieved by other techniques. The precise role of *ras* oncogenes in human tumors remains obscure; however, it seems likely that these genes play a direct causal role in the development of these diseases, and that the rapid and sensitive identification of tumors containing *ras* mutations will have real clinical value as our understanding of their action increases.

REFERENCES

1. Barbacid, M. (1987) Ann. Rev. Biochem. 56:779-827.
2. Bos, J.L., Verlaan-de Vries, M. Janssen, A.M., Veeneman, G.H., van Boom, J.H., and van der Eb, A.J. (1984) *Nucl. Acids Res.* 12:9155-9163.
3. Verlaan-de Vries, M., Bogaard, M.E., van den Elst, H., van Boom, J.H., van der Eb, A.J., and Bos, J.L. (1986) *Gene* 50:313-320.
4. Bos, J.L., Fearon, E.R., Hamilton, S.R., Verlaan-de Vries, M., van Boom, J.H., van der Eb, A.J., and Volgelstein, B. (1987) *Nature* 327:293-297.
5. Shibata, D., Arnheim, N., and Martin, J. (1988) *J. Exp. Med.* 167:225-230.
6. Saiki, R., Sharf, S., Faloona, F., Mullis, K., Horn, G., Ehrlich, H.A., and Arnheim, N. (1985) *Science* 230:1350-1353.
7. Winter, E., Yamamoto, F., Almoguera, C., and Perucho, M. (1985) *Proc. Natl. Acad. Sci. USA* 82:7575-7579.
8. Almoguera, C., Shibata, D., Forrester, K., Martin, J., Arnheim, N., and Perucho, M. (1988) *Cell* 53:549-554.
9. McMahon, G., Davis, E., and Wogan, G.N. (1987) *Proc. Natl. Acad. Sci. USA* 84:4974-4978.
10. Collins, S.J. (1988) *Mutat. Res.* 195:255-271.
11. Bos, J.L. (1988) *Mutat. Res.* 195:255-271.

12. Smit, V.T.H.B.M., Boot, A.J.M., Smits, A.M.M., Fleuren, G.J., Cornelisse, C.J., and Bos, J.L. (1988) *Nucl. Acids Res.* 16:7773-7782.
13. Qualfe, C.J., Pinkert, C.A. Ornitz, D.M., Palmiter, R.D., and Brinster, R.L. (1987) *Cell* 48:1032-1034.
14. Vogelstein, B., Fearon, E.R., Hamilton, S.R., Kern, S.E., Preisinger, A.C., Leppert, M., Nakamura, Y., Whyte, R., Smits, A.M.M., and Bos, J.L. (1988) 319:525-532.
15. Forrester, K., Almoguera, C., Han, K., Grizzle, W.E., and Perucho, M. (1987) *Nature* 327:298-303.
16. Rodenhuis, S., van de Wetering, M.L., Mooi, W.J., Evers, S.G., van Zandwijk, N., and Bos, J.L. (1987) *N. Engl. J. Med.* 317:929-935.
17. Rodenhuis, S., Slebos, R.J., Boot, A.J.M., Evers, S.G., Mooi, W.J., Wagenaar, S.S., van Bodegom, P.C., and Bos, J.L. (1988) Submitted.
18. Lemoine, N.R., Mayall, E.S., Wyllie, F.W., Dillwyn-Williams, E., Goyns, M., Stringer, B., and Wynford-Thomas, D. (1989) in press.
19. Suarez, H.G., Du Villard, J.A., Caillou, B., Schlumberger, M., Tubiana, M., Parmentier, C., and Monier, R. (1988) *Oncogene* 2:403-406.
20. Fusco, A., Grieco, M., Santoro, Berlingieri, M.J., Pilotti, S., Pierotti, M.A., Dell Porta, G., and Vecchio, G. (1987) *Nature* 328:170-172.
21. Raybaud, F., Noguchi, T., Marics, I., Adelaide, J., Planche, J., Batoz, M., Aubert, C., de Lapeyriere, O., and Birnbaum, D. (1988) *Cancer Res.* 48:950-953.
22. Van't Veer, L.J., Burgering, B.M.Th., Versteeg, R., Boot, A.J.M., Ruiter, D.J., Osanto, S., Schrier, P.I., and Bos, J.L. (1989) Submitted.
23. Raskind, W.H., Triumali, N., Jacobson, R., Singer, J., and Fialkow, P.J. (1984) *Blood* 63:1318-1323.
24. Nowell, P.C. (1976) *Science* 194:23-28.
25. Bos, J.L., Toksoz, D., Marshall, C.J., Verlaan-de Vries, M., Veeneman, G.H., van der Eb, A.J., van Boom, J.H., Janssen, J.W.G., and Steenvoorden, A.C.M. (1985) *Nature* 315:726-730.
26. Bos, J.L., Verlaan-de Vries, M., van der Eb, A.J., Janssen, J.W.G., Delwel, R., Lowenberg, B., and Colly, L.P. (1987) *Blood* 69:1237-1241.
27. Needleman, S.W., Kraus, M.H., Srivastava, S.K., Levine, P.H., and Aaronson, S.A. (1986) *Blood* 67:753-757.
28. Hirai, H., Kobayashi, Y, Mano, H., Hagiwara, K., Maru, Y., Omine, M., Mizoguchi, H., Nishida, J., and Takaku, F. (1987) *Nature* 327:430-432.
29. Janssen, J.W.G., Steenvoorden, A.C.M., Lyons, J., Anger, B., Bohlke, J.U., Bos, J.L., Seliger, H., and Bartram, C.R. (1987) *Proc. Natl. Acad. Sci. USA* USA 84:9228-9232.
30. Lui, E., Hjelle, B., Morgan, R., Hecht, F., and Bishop, J.M. (1987) *Nature* 330:186-188.
31. Toksoz, D., Farr, C.J., and Marshall, C.J. (1987) *Oncogene* 1:409-413.
32. Farr, C.J., Saiki, R.K., Ehrlich, H.A., McCormick, F., and Marshall, C.J. *Proc. Natl. Acad. Sci. USA* 85:1629-1633.
33. Lyons, J., Janssen, J.W.G., Bartram, C.R., Layton, D.M., and Mufti, G.J. (1988) *Blood* 71:1707-1711.
34. Senn, H.P., Fopp, J.M., Schmid, L., and Moroni, C. (1988) *Int. J. Cancer* 41:59-64.
35. Padua, R.A., Carter, G., Hughes, D., Gow, J., Farr, C., Oscier, D., McCormick, F., and Jacobs, A. (1988) *Leukemia* 2:503-510.
36. Yunis, J.J., Boot, A.J.M., Mayer, M.G., and Bos, J.L. (1988) Submitted.
37. Senn, H.P., Fopp, J.M., Schmid, L., and Moroni, C. (1988) *Blood* 72:931-935.

38. Kraus, M.H., Yuasa, Y., and Aaronson, S.A. (1984) *Proc. Natl. Acad. Sci, USA* 81:5384-5388.
39. Van de Vijver, M., Peterse, J.L., Mooi, W.J., Lohmans, J., Verbruggen, M., van de Bersselaar, R., Devilee, P., Cornelisse, C., Bos, J.L., Yarnolad, J., and Nusse, R. (1988) *Cancer Cell* in press.
40. Spandidos, D.A. (1987) *Anticancer Res.* 7:991-996.
41. Van't Veer, L.J., Hermens, R., van den Berg-Bakker, L.A.M., Cheng, N.Ching, Fleuren, G.J., Bos, J.L., Cleton, F.J., and Schrier, P.I. (1988) *Oncogene* 2:157-165.
42. Hollstein, M.C., Smits, A.M., Galiana, C., Yamasaki, H., Bos, J.L., Mandard, A., Partensky, C., and Montesano, R. (1988) Submitted.
43. Ballas, K., Lyons, J., Janssen, J.W.G., and Bartram, C.R. (1988) *Eur. J. Ped.* 147:313-314.
44. Sakato, H., Mori, M., Taira, M., Yoshida, T., Matsukawa, S., Shimizu, K., Sekiguchi, M., Terada, M., and Sugimura, T. (1986) *Proc. Natl. Acad. Sci. USA* 83:3997-4001.
45. Fujita, K.N., Ohuchi, T. Yao, Okumura, M., Fukushima, Y., Kanakura, V., Kitamura, Y., and Fujita, J. (1987) *Gastroenterology* 6:1339-1345.
46. Fujita, J., Kraus, M.H., Onoue, H., Srivastava, S.K., Ebi, Y., Kitamura, Y., and Rhim, J.S. (1988) *Cancer Res.* 48:5251-5255.
47. Leon, J., Kamino, H., Steinberg, J.J., and Pellicer, A. (1988) *Mol. Cell. Biol.* 8:786-793.
48. Fujita, J., Srivastava, S.K., Kraus, M.H., Rhim, J.S., Tronick, S.R., and Aaronson, S.A. (1985) *Proc. Natl. Acad. Sci. USA* 82:3849-3853.
49. Malone, P.R., Visvanathan, K.V., Ponder, B.A., Shearer, R.J., and Summerhayes, Il.C. (1985) *Br. J. Urol.* 57:664-667.
50. Visvanathan, K.V., Pocock, R.D., and Summerhays, I.C. (1988) *Oncogene Res.* 3:77-86.
51. Liu, E., Hjelle, B., and Bishop, J.M. (1988) *Proc. Natl. Acad. Sci. USA* 85:1952-1956.
52. Rodenhuis, S., Bos, J.L., Slater, R.M., Behrendt, H., Van't Veer, M., and Smets, L.A. (1986) *Blood* 67:1698-1704.
53. Neri, A., Knowles, D.M., Greco, A, McCormick, F., and Dalla-Favera, R. (1988) *Proc. Natl. Acad. Sci. USA* 85:9268-9272

CHAPTER 19

Application of PCR to the Detection of Human Infectious Diseases

Shirley Kwok and John J. Sninsky

INTRODUCTION

The polymerase chain reaction (PCR)[1,2] has revolutionized the detection of bacterial, fungal, and viral pathogens. The targeted amplification of nucleic acid sequences provides not only dramatic increases in the number of copies to be detected but concomitantly provides a nearly equivalent reduction in the complexity of the nucleic acid to be probed. Either DNA or RNA (following the production of complementary DNA using reverse transcriptase) can be used as a template for amplification. Further, since PCR is a target rather than a signal amplification system, the benefit of the variety of procedures developed over the years to amplify signal can be exploited coincidentally. Combined, these aspects of PCR allow ready detection of single pathogenic organisms or virus particles, an accomplishment provided earlier by procedures employing the *in vitro* propagation of such pathogens. However, since the exponential amplification of PCR is catalyzed by a biochemically simple cyclical process requiring less than minutes per cycle, this procedure promises to supplant the culturing of a pathogen which frequently requires a total elapsed time of days to weeks. By definition, pathogens not capable of *in vitro* propagation, for example because of the inability to culture a specific host cell for a virus, are refractory to detection using this approach. Polyclonal and monoclonal antibodies used in a diagnostic setting typically recognize haptens found in multiple copies on a pathogen to circumvent the need to replicate the desired pathogen. Unfortunately, the cross-reactivity of these antibodies with host haptens and related but medically distinct pathogens as well as their often low avidity has compromised the convenient and broader use of these diagnostic reagents. In addition, some viruses establish latent infections in which active viral replication is substantially attenuated thereby obviating procedures requiring the detection of proteins.

Although most diagnostic assays are based on sample collection by non-invasive means, invasive sampling is sometimes unavoidable. By definition, optimal invasive sampling requires the collection of minuscule quantities of material. Even needle punch biopsies and aspirates of various types provide sufficient material for PCR analysis.

This chapter will discuss the procedure, in general, in terms of its application to the infectious disease diagnostic arena, highlight the recent successful application of PCR to the detection of various pathogens with an emphasis on viruses, and point out the likely direction of future studies in this area.

PROCEDURE

Analytical Sensitivity and Specificity

The analytical sensitivity of PCR is dependent on the specificity at which the primers initiate polymerization from the targeted site relative to other non-specific regions of nucleic acid. Altering the concentration of the various reaction components such as the enzyme, primers, and metal cation (magnesium chloride) has a marked affect on the yield of the desired PCR product.[3,4] Additionally, the temperatures and times of the annealing and extension steps of each cycle also affect sensitivity and specificity of the amplifications.[5] Although primers of 20 to 30 residues are commonly used, synthetic oligonucleotides containing as few as twelve bases have been utilized.[6] Since pathogen detection is often carried out in the presence of vast quantities of host nucleic acid, the ability of the primers to be pathogen-specific is critical to sensitive and specific amplifications. The size of the human genome, is such that if it contained random sequences, a particular sequence would need to encompass sixteen residues to be represented only once. Of course, since the human genome is not random, regions containing a specific sequence greater than sixteen bases may be represented thousands of times and, correspondingly, sequences shorter than sixteen bases may be unique in the genome. Until a larger proportion of the sequence of the human genome has been determined, the design of optimal primers for specific pathogens will require significant empirical experimentation. It should be noted, however, that in the systems best optimized to date (i.e., globin[3] and HIV[7,8]) the detection of one to ten molecules of DNA in the context of highly complex nucleic acid from the equivalent of one million cells can be accomplished using thirty cycles of amplification and standard detection strategies. Of course, as the number of copies of a nucleic acid that one wishes to detect in a sample approaches one, the reproducibility of detection is compromised because of the decreased likelihood that all aliquots of the sample will contain the desired target and stochastic effects of the amplification reaction.

The specific detection of human retroviruses merits further comment. Following cell entry, the retroviral life cycle includes conversion of the single-stranded RNA genomes to circular double-stranded proviral DNA which subsequently covalently integrates into the host's cellular genome. The proviral genome remains integrated for the life of the cell. The infection of germ line cells results to the eventual vertical transmission of the viral sequence to the progeny of the infected host. Past studies (see ref. 9 for a review) of the endogenous viral sequences of the human genome have led to several conclusions: 1) multiple families reside in the genome, 2) family members vary in sequence and structure, 3) nearly intact as well as substantially altered members are present, 4) numerous, non-tandem copies are present, 5) the endogenous sequences resulted from infection post-speciation, and 6) they may share significant sequence homology with pathogenic exogenous retroviruses. The role, if any, that these sequences play in the development of human disease is unclear at this time. To date, it appears as though the retroviral subfamily *Oncovirinae* have participated in the evolution of the human genome to a greater extent than either the *Lentivirinae* or the Hepadnaviruses (a class of viruses that similar to retroviruses, replicate through an RNA intermediate using a virally encoded reverse transcriptase but package DNA rather than RNA; prototype, hepatitis B virus (HBV)) but additional studies are required to resolve this issue. Since thousands of copies of these retroviral remnants are present, the potential amplification of these sequences must be considered when conserved regions of exogenous retroviral sequences are targeted. Specifically, amplification of these elements may lead to false positives (perhaps sporadic given stochastic initiation of polymerization) or to compromised sensitivity of the desired viral sequence by competing for the primers or enzyme.

Viral Heterogeneity

Viral genomes, particularly those of RNA viruses and viruses that replicate through an RNA intermediate (i.e., retroviruses and hepadnaviruses) contain multiple base alterations, additions, duplications, and deletions.[10] The variability of these viruses has been attributed to the low fidelity and lack of proofreading functions of the polymerases responsible for replication. In addition, since RNA polymerase II plays a critical role in the retroviral and hepadnaviral life cycles, both the fidelity of this enzyme and the reverse transcriptase must be taken into consideration. Repeated rounds of infection further magnify variability. The role that these viral variants play in the natural history of infection is only beginning to be understood and appears to vary with each class of virus. This variability may lead to problems when using PCR since some individuals may harbor viral variants incapable of amplification due to sequence alterations in the region recognized by the primers. As a result, continued evaluation of the clinical sensitivity of a primer pair and probe system and the development of primers that may tolerate mismatches more efficiently should be an integral part of the evaluation of a diagnostic assay for these types of viruses. The role that mismatches between target and primer play in efficient primer extension is beginning to undergo more extensive study.[11-13] Not surprisingly, many of these investigations have begun to evaluate the effect of single 3' terminal mismatches. The efficiency of extension of a primer:target duplex mismatched at the 3' terminal base is expected to depend to a great extent on the reaction parameters (i.e., concentrations of metal cation, primers and enzyme). However, preliminary results suggest that some mismatches may allow relatively efficient amplification.

Carryover

Although extraneous nucleic acid from multiple sources may serve as a template for amplification thereby leading to false positive reactions, the predominant source appears to be the PCR product from previous reactions. Caution, therefore, should be used when numerous amplifications of the same primer pair system are being used.[14] The implications of false positives in the context of HIV-1 diagnostic assays prompted us to develop a series of recommendations which should be followed in order to reduce the frequency and amount of carryover.[15] *First,* there should be a physical separation of reactions prior to and following amplification. In addition, separate sets of devices such as automatic pipettors, disposable pipettes, microfuge tubes, and gloves should be kept in each area. *Second,* positive displacement pipettors should be used. *Third,* reaction reagents should be aliquoted so as to minimize the number of times a particular stock solution is used. *Fourth,* laboratory precautions typical of handling radioactivity should be incorporated. *Fifth,* the number and type of controls should be chosen judiciously. Specifically, there should be a small number of positive controls with a low target molecule concentration, a large number of both reagent controls (i.e., reactions lacking only template nucleic acid) and pathogen negative controls. A dramatic decrease in observed carryover will result if these suggestions are heeded.

Clinical Utility

The clinical utility of any diagnostic assay is based on its clinical sensitivity and specificity when carried out in low risk and high risk populations.

Often times, the predictive value of the assay is critically linked to the setting of the background cutoff level. PCR-based assays, not unlike other technologies, will also require considerable clinical trials with statistically significant numbers of samples to determine the optimal background subtraction levels. Since PCR employs amplification, the number of cycles or degree of amplification will need to be adjusted on an assay by assay basis. There will be diagnostic targets such as the human immunodeficiency virus (HIV) that will likely require extensive amplification and perhaps others such as Epstein Barr Virus (EBV) or Cytomegalovirus (CMV) that because of their prevalence in a significant fraction of unaffected individuals may require lower levels of amplification to avoid "false positives." In addition, studies in low versus high risk populations may demonstrate greater predictive value in one group versus another.

VIRUSES

Human Retroviruses

The human retroviruses characterized to date (human immunodeficiency virus types 1 (HIV-1)[16,17] and 2 (HIV-2)[18] and human T cell lymphotrophic virus types I (HTLV-I)[19] and II (HTLV-II)[21,22]) and those yet to be identified are associated with, responsible for, or implicated in an ever increasing number of human diseases and disorders. Although serological assays for antibodies to these viruses have played a significant role in the identification of individuals who are infected with these viruses, direct detection of the viral pathogen is highly desirable in some cases. Complicating factors associated with retroviral detection include: 1) the transcriptional dormancy of the proviral genome, 2) the relatively small number of infected cells within the peripheral blood,[22,23] 3) the small number of proviral copies per infected cell, 4) the existence of multiple related but distinct viral members, and 5) the presence of infected cells in reservoirs such as the brain and lymph node that are not readily amenable to monitoring.[23] PCR has proven to be an ideal procedure for the detection of these viruses.

HIV, Types 1 and 2

HIV-1 and HIV-2 are members of the *Lentivirnae* subfamily of retroviruses and are responsible for the development of Acquired Immunodeficiency Disease Syndrome (AIDS) and numerous associated opportunistic infections. The two viruses share approximately 60% overall nucleic acid sequence homology for the *gag* and *pol* genes and 30-40% for the other viral genes and LTR. Several procedures have been developed for the direct detection of these viruses including the *in vitro* propagation of virus and the use of reverse transcriptase or viral nucleocapsid antigen (p24) assays to monitor the cultures, and the use of the nucleocapsid antigen assay for the direct detection of virus in peripheral blood.[24]   The successful culturing of virus from seropositive individuals and those with documented infection varies widely.   Moreover, the procedure is labor intensive, time consuming, and costly.   The use of the direct viral antigen assay for blood suffers from the lack of sensitivity, presumably due to the establishment of a latent infection, and is compromised by the sequestering of viral antigen in circulating immune complexes.   Other procedures including *in situ* hybridization[25] and Southern blot analysis[23] have been used with sporadic success because of the small number of infected peripheral blood mononuclear cells.

The initial application of PCR to the detection of HIV-1 targeted DNA as a template so as to provide detection of infected cells both actively replicating virus as well as those that were transcriptionally dormant.[26] Limited studies[27,28] using RNA as a template for amplification have proven less sensitive than when DNA is targeted for PCR but since active transcription may potentially provide larger numbers of copies, this approach should continue to be evaluated.   Given the extensive sequence heterogeneity of the HIV-1 viral genome,[29-31] it was necessary to amplify only highly conserved regions.   The availability of nucleic acid alignments of all sequenced HIV isolates will expedite the identification of conserved regions.

PCR has demonstrated clinical utility for identifying HIV proviral sequences in 1) infected individuals prior to the generation of antibodies, the so-called seronegative window,[32] 2) resolving the infection status of individuals with ambiguous or indeterminate serological status using the Western blot assay, 3) documenting infection in seropositive individuals negative by other direct detection assays,[8] 4) screening of neonates,[33,34] and 5) determining the type of virus present.[35]   The development of more rapid, colorimetric, and inexpensive PCR assays will make this approach more amenable to large scale screening.

HTLV, Types I and II

HTLV-I is a member of the *Oncovirinae* subfamily of retroviruses and represents the first human retrovirus characterized.[19]   This virus is associated with the development of an unusually aggressive type of leukemia referred to as adult T cell leukemia (ATL).   Not long thereafter, another virus, HTLV-II, sharing about 60% overall nucleic acid sequence homology, was isolated from individuals with a rare T cell variant Hairy cell leukemia.[20,21] Subsequent studies have been unable to substantiate this proposed linkage. After a protracted period, serological assays for HTLV-I were approved and mandated for blood bank screening.   Virus culture has played a seminal role in the direct detection of this virus from infected individuals.   Several other procedures for direct detection of this pathogen, similar to those noted above, have been used but all appear to have compromised sensitivity and specificity.

An association of HTLV-I and a chronic progressive myelopathy characterized by degenerative lower limb paralysis termed tropical spastic paraparesis (TSP) in the Caribbean and HTLV-I associated myelopathy (HAM) in temperate regions was proposed because of the cross-reactivity of antibodies in affected individuals and HTLV-I viral lysates. Final resolution of this association required the critical demonstration that prototypical HTLV-I rather than a related virus was present. PCR played a key role in this demonstration by unequivocal detection of HTLV-I in a significant number of patients and by providing sufficient material to molecularly clone and sequence a region of HTLV-I from one of these patients.[36,37]

PCR has demonstrated clinical utility for the detection of the HTLV family for 1) identifying infected individuals prior to seroconversion,[38] 2) confirming infection in seropositive individuals, 3) documenting infection in individuals presenting with symptoms dissimilar from classic ATL,[39] and 4) typing the virus present (i.e., type I versus type II).[37,38] The latter value is increasingly important given the near epidemic prevalence of HTLV-II in the intravenous drug community[40,41] and the inability, to date, to link this virus with a specific disease, if any.

Hepatitis B Virus

PCR has dramatically increased the sensitivity of detection of hepatitis B virus. Several laboratories[42,43] have reported the detection of as few as three HBV genomes (or about 300 virus particles per milliter of serum) which represents a 10,000-fold increase in sensitivity over standard procedures. In other studies, HBV DNA has been detected in the serum of 7 of 382 (1.8%) healthy HBsAg negative blood donors in China[44] using slot blot hybridization. In addition, PCR has been used to identify and characterize HBV DNA in three patients negative for all HBV serological markers.[45]

Cytomegalovirus

Cytomegalovirus (CMV) is an important pathogen in immunocompromised patients and newborns. Congenital viral infection can lead to mental retardation and nonhereditary sensorineural deafness. The advent of chemotherapy for CMV infections makes early and sensitive detection of this virus particularly important. Investigators have demonstrated clinical utility in using PCR to detect CMV in urine specimens of infected infants[46] and in AIDS patients[47] with greater sensitivity than virus culture.

Human Papillomavirus

The genital human papillomaviruses (HPVs) are a diverse group of distinct virus types associated with a number of diseases and cancers (for example, see ref. 48). Distinct genital HPV types have been linked to different clinical manifestations. For example, types 16 and 18 are found in cervical dysplasia and carcinoma; types 6 and 11 are associated with benign condylomas. The detection and typing of genital HPVs in normal and diseased tissue samples will be critical to understanding the role they play in the development of various diseases and cancer. The PCR-based assays described to date[49,50] suggest that this procedure will prove unusually valuable as a sensitive and specific tool.

FUTURE APPLICATIONS

The pathogens characterized to date will represent only a fraction of the agents responsible for human disease. The identification of human retroviruses and their association with disease, in particular, is in its infancy. Since related viruses share short hyphenated regions of homology, PCR is ideally suited to facilitate the detection of uncharacterized viruses. The conserved regions of the genes responsible for replication and the regulation of the viral life cycle serve as productive targets for PCR. In model studies, coincident detection of members of Oncoviruses[38] and hepadnaviruses[6] was achieved with primers to one of these conserved regions. The development of procedures to convert PCR to a quantitative assay has begun and promises to dramatically assist in the evaluation of therapeutic trials for antiviral intervention. Although published studies have yet to appear describing the application of PCR to the detection of medically relevant bacteria and fungi, the numerous reports that have appeared in recent meetings suggests that these areas of human diagnostics will soon yield to this powerful technique. Likewise, the use of PCR in veterinary diagnostics will also gain acceptance in the not too distant future.

CONCLUSION

The use of PCR, although initially used for persistent and latent viruses such as the human retroviruses, will gain broader acceptance in the diagnostic community as a whole. This *in vitro* DNA amplification procedure will be used initially in clinical reference laboratories because of the expense and sophisticated nature of the present assays but upon the development of simpler and inexpensive formats will be exploited as diagnostic kits.

REFERENCES

1. Mullis, K.B., and Faloona, F.A. (1987) *Meth. Enzymol.* 155:335-50.
2. Saiki, R.K., Scharf, S., Faloona, F., Mullis, K.B., Horn, G.T., Erlich, H.A., and Arnheim, N. (1985) *Science* 230:1350-1354.
3. Saiki, R.K., Gelfand, D.H., Stoffel, S., Scharf, S.J., Higuchi, R., Horn, G.T., Mullis, K.B., and Erlich, H.A. (1988) *Science* 239:487-491.
4. Kwok, S., Mack, D.H., Kellogg, D.M., McKinney, N., Faloona, F., and Sninsky, J.J. (1989) *Current Communications in Molecular Biology: Polymerase Chain Reaction*. Erlich, H.A., Gibbs, R., and Kazazian, H.H., Jr., eds. Cold Springs Harbor Laboratories, in press.

5. Saiki, R. (1989) Chapter 1, this book.
6. Mack, D.H., and Sninsky, J.J. (1988) *Proc. Natl. Acad. Sci. USA* 85:6977-6981.
7. Kwok, S., Mack, D.H., Sninsky, J.J., *et al.* (1989) *HIV Detection by Genetic Engineering Methods.* Luciw, P.A., Steimer, , K.S., eds. Marcel Dekker, Inc., New York and Basel, pp. 243-253.
8. Ou, C.Y., Kwok, S., Mitchell, S.W., Mack, D.H., Sninsky, J.J., Krebs, J.W., Feorino, P., Warfield, D., and Schochetman, G. (1988) *Science* 239:295-297.
9. Shih, A., Misra, R., and Rush, M.G. (1989) *J. Virol.* 63:64-75.
10. Steinhauer, D.A., and Holland, J.J. (1986) *Ann. Rev. Microbiol.* 41:409-433.
11. Aboulela, F., Koh, D., Tinoco, I., and Martin, F.H. (1985) *Nucl. Acids Res.* 13:4811-4824.
12. Petruska, J., Goodman, M.F., Boosalis, M.S., Sowers, L.C., Cheong, C., and Tinoco, I. (1988) *Proc. Natl. Acad. Sci. USA* 85:6252-6256.
13. Newton, C.R., Graham, A., Heptinstall, L.E., Powell, S.J., Summers, C., Kalsheker, N., Smith, J.C., and Markham, A.F. (1989) *Nucl. Acids Res.* 17:2503-2516.
14. Schochetman, G., Ou, G.Y., and Jones, W.K. (1988) *J. Infec. Dis.* 158:1154-1157.
15. Kwok, S., and Higuchi, R. (1989) *Nature* 339:237.
16. Barre-Sinoussi, F., Chermann, J.C., Rey, F., Nugeyre, M.T., Chamaret, S., Gruest, J., Dauguet, C., Axler-Blin, C., Vezinet-Brun, F., Rouzioux, C., Rozenbaum, W., and Montagnier, L. (1983) *Science* 220:868-870.
17. Popovic, M., Sarngadharan, M.G., Read, E., and Gallo, R.C. (1984) *Science* 224:497-500.
18. Guyader, M., Emerman, M., Sonigo, P., Clavel, F., Montagnier, L., and Alizon, M. (1987) *Nature* 326:662-669.
19. Poiesz, B.J., *et al.* (1980) *Proc. Natl. Acad. Sci. USA* 77:7415.
20. Kalyanaraman, V.S., Sarngadharan, M.G., Robert-Guroff, M., Miyoshi, I., Blayney, D., Golde, D., and Gallo, R.C. (1982) *Science* 218:571.
21. Rosenblatt, J.D., Golde, D.W., Wachsman, W., Giorgi, J.V., Jacobs, A., Schmidt, G.M., Quan, S., Gasson, J.C., and Chen, I. (1986) *N. Engl. J. Med.* 315:72.
22. Shaw, G.M., Hahn, B.H., Arya, S.K., Groopman, J.E., Gallop, R.C., and Wong-Staal, F. (1984) *Science* 226:1165-1170.
23. Shaw, G.M., Harper, M.E., Hahn, B.H., *et al.* (1985) *Science* 177-181.
24. Wittek, A.E., Phelan, M.A., Well, M.A., Vujcic, L.K., Epstein, J.S., Lane, C., and Quinnan, G.V. (1987) *Ann. Int. Med.* 107:286-292.
25. Harper, M.H., Marselle, L.M., Gallo, R.C., and Wong-Staal, F. (1986) *Proc. Natl. Acad. Sci. USA* 83:772-776.
26. Kwok, S., Mack, D.H., Mullis, K.B., Poiesz, B., Ehrlich, G., Blair, D., Friedman-Kien, A., and Sninsky, J.J. (1987) *J. Virol.* 61:1690-1694.
27. Byrne, B.C., Li, J.J., Sninsky, J., and Poiesz, B.J. (1988) *Nucl. Acids Res.* 16:4165.
28. Hart, C., Schochetman, G., Spira, T., Lifson, A., Moore, J., Galphi, M., Sninsky, J., and Ou, C.Y. (1988) *Lancet* 2:596-599.
29. Shaw, G.M., Hahn, B.H., Arya, S.K., Groopman, J.E., Gallo, R.C., and Wong-Staal, F. (1984) *Science* 226:1165-1171.
30. Saag, M.S., Hahn, B.H., Gibbons, J., Li, Y., Parks, E.S., Parks, W.P., and Shaw, G.M. (1988) *Nature* 334:440-444.
31. Goodenow, M., Huet, T., Saurin, W., Kwok, S., Sninsky, J., and Wain-Hobson, S. (1989) *J. AIDS*, in press.
32. Imagawa, D.T., Lee, M.H., Wolinsky, S.M., *et al.* (1989) *N. Engl. J. Med.* 320:1458-1462.

33. Laure, F., Courgnaud, V., Rouzioux, C., Blanche, S., Veber, F., Burgard, M., Jacomet, C., Griscelli, C., and Brechot, C. (1988) *Lancet* 2:538-541.
34. DeRossi, A., Amdori, A., Chieco-Bianchi, L., *et al.* (1988) *Lancet* 2:278.
35. Rayfield, M., DeCock, K., Heyward, W., Goldstein, L., Krebs, J., Kwok, S., Lee, S., McCormick, J., Moreau, M.M., Odehouri, K., Schochetman, G., Sninsky, J., and Ou, C.-Y. (1988) *J. Infect. Dis.* 158:1170-1176.
36. Bhagavati, S., *et al.* (1988) *N. Engl. J. Med.* 318:1141-1147.
37. Kwok, S., Kellogg, D., Ehrlich, G., Poiesz, B., Bhagavati, S., and Sninsky, J.J. (1989) *J. Infec. Dis.* 158:1193-1197.
38. Kwok, S., Ehrlich, G., Poiesz, B., Kalish, R., and Sninsky, J.J. (1988) *Blood* 72:1117-1123.
39. Duggan, D.B., Ehrlich, G.D., Davey, F.P., Kwok, S., Sninsky, J., Goldberg, J., Baltrucki, L., and Poiesz, B. (1988) *Blood* 71:1027-1032.
40. Lee, H., Swanson, P., Shorty, V.S., Zack, J., Rosenblatt, J.D., and Chen, I.S-Y (1989) *Science* 244:471-475.
41. Ehrlich, G., *et al.* (1989) *Blood*, in press.
42. Larzul, D., Guigue, F., Sninsky, J.J., Mack, D.H., Brechot, C., and Guesdon, J.L. (1988) *J. Virol. Methods* 20:227-237.
43. Kaneko, S., Miller, R.H., Feinstone, S.M., Unoura, M., Kobayashi, K., Hattori, N., and Purcell, R.H. (1989) *Proc. Natl. Acad. Sci. USA* 86:312-316.
44. Sun, C.-F., Pao, C., Wu, S.-Y., and Liaw, Y.-F. (1988) *J. Clin. Microbiol.* 1848-1852.
45. Thiers, V., Nakahima, E., Kremsdorf, D., Mack, D., Schellekens, H., Driss, F., Goudeau, A., Wands, J., Sninsky, J., and Tiollais, P. (1988) *Lancet* 2:1273-1276.
46. Demmler, G.J., Buffone, G.J., Schimbor, S.M., and May, R.A. (1988) *J. Infec. Dis.* 158:1177-1184.
47. Shibata, D., Martin, W.J., Appleman, M.D., *et al.* (1988) *J. Infec. Dis.* 158:1185-1192.
48. Broker, T.R., and Botchan, M. (1986) *Cancer Cells* 4:17.
49. Shibata, D., Arnheim, N., and Martin, W.J. (1988) *J. Exp. Med.* 167:225.
50. Manos, M.M., Ting, Y., Wright, D.K., Lewis, A.J., Broker, T.R., and Wolinsky, S.M. (1989) *Cancer Cells* 7:209-214.

Index